WELCOME TO LINDAL

There's a special joy in living in a home of your own design. At Lindal, the pleasures begin with the planning. And the planning begins with this idea book, filled with all the information and inspiration you need to create the  home of your dreams. So make yourself at home. Browse through our idea book. And let your imagination soar. Because your Lindal custom cedar home truly can be anything—and everything— you want it to be.

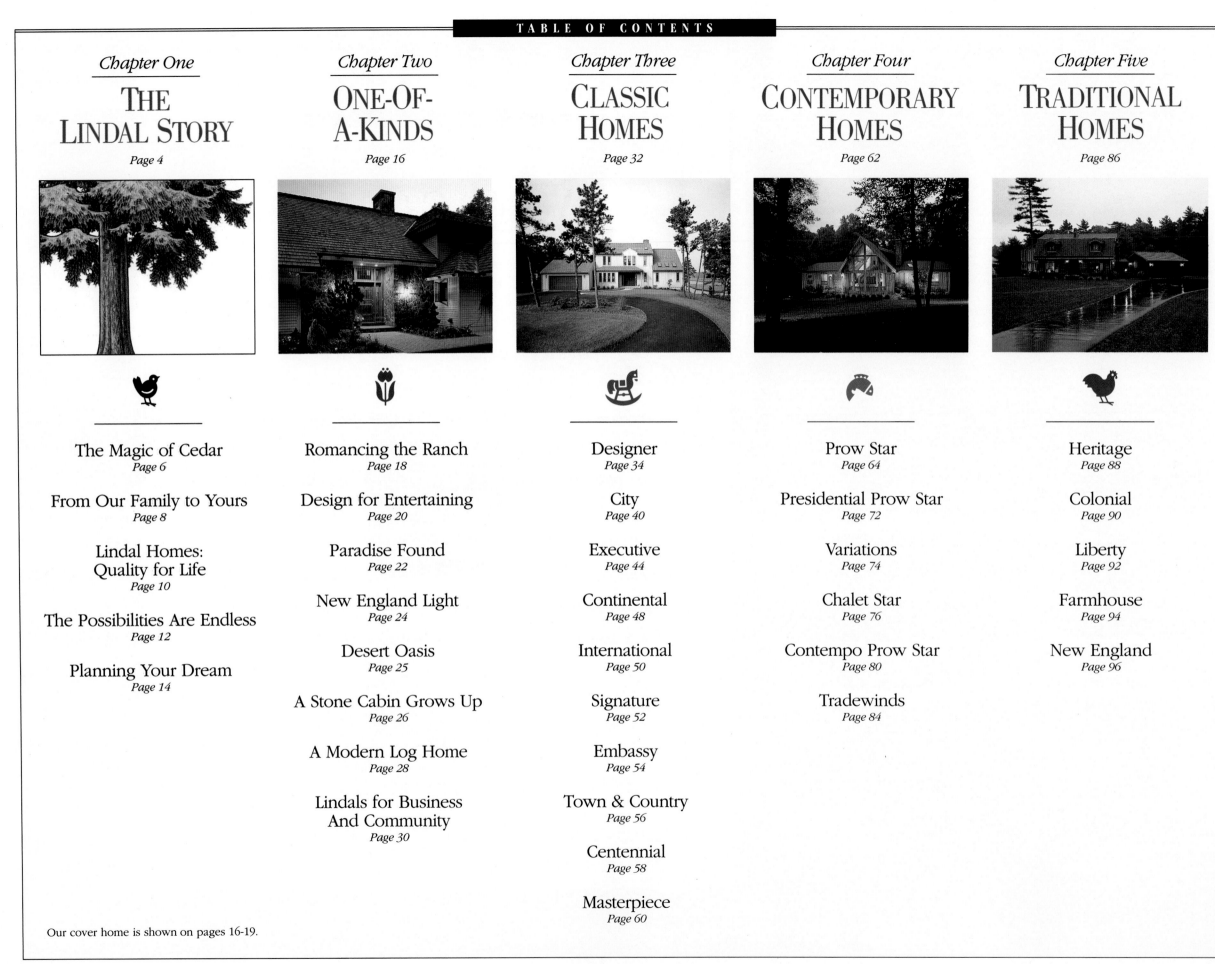

Chapter One

THE LINDAL STORY
Page 4

The Magic of Cedar
Page 6

From Our Family to Yours
Page 8

Lindal Homes:
Quality for Life
Page 10

The Possibilities Are Endless
Page 12

Planning Your Dream
Page 14

Chapter Two

ONE-OF-A-KINDS
Page 16

Romancing the Ranch
Page 18

Design for Entertaining
Page 20

Paradise Found
Page 22

New England Light
Page 24

Desert Oasis
Page 25

A Stone Cabin Grows Up
Page 26

A Modern Log Home
Page 28

Lindals for Business
And Community
Page 30

Chapter Three

CLASSIC HOMES
Page 32

Designer
Page 34

City
Page 40

Executive
Page 44

Continental
Page 48

International
Page 50

Signature
Page 52

Embassy
Page 54

Town & Country
Page 56

Centennial
Page 58

Masterpiece
Page 60

Chapter Four

CONTEMPORARY HOMES
Page 62

Prow Star
Page 64

Presidential Prow Star
Page 72

Variations
Page 74

Chalet Star
Page 76

Contempo Prow Star
Page 80

Tradewinds
Page 84

Chapter Five

TRADITIONAL HOMES
Page 86

Heritage
Page 88

Colonial
Page 90

Liberty
Page 92

Farmhouse
Page 94

New England
Page 96

Our cover home is shown on pages 16-19.

Chapter Six
SMALL TREASURES
Page 98

Chapter Seven
HOME PLANNING IDEAS
Page 110

Chapter Eight
SPECIFICATIONS
Page 130

Chapter Nine
PLANS
Page 146

Chapter Ten
HOW TO GET YOUR DREAM HOME
Page 226

Prow
Page 100

Chalet
Page 101

Summit
Page 102

Contempo Prow
Page 104

View
Page 105

Panorama
Page 106

Gambrel
Page 108

Pole
Page 109

Before You Start
Page 112

From Site to Floorplan
Page 114

Kitchen Success
Page 116

The Bathroom:
From Basic to Luxurious
Page 120

Windows:
Designing for Light
Page 122

Eye-Catching Openings
Page 124

Hardwood Floors, Decks,
Lanais, Screened Porches
Page 126

Sunrooms Et Cetera
Page 128

Lindal Means Quality
Page 132

Post & Beam
Page 134

Our Four Major Home Styles
Page 136

Strong, Efficient Building
Systems – From the
Ground Up
Page 138

Going Solar
Page 144

How to Personalize A Plan
Page 148

Reading the Plans
Page 150

Lindal Planning Aids
Page 152

Planning Grid
Fold-Out

Plans
Page 153

Plans Index
Page 224

Here's What Your Dealer
Can Do for You
Page 228

There's A Lot to Love About
Your Lindal
Page 230

What Will Your Lindal Cost?
Page 230

From Our Family to Yours
Page 232

Index
Inside Back Cover

*Library of Congress Number
89-063728*

CHAPTER ONE

THE LINDAL STORY

Building your own home is one of the most important investments you'll ever make. At Lindal Cedar Homes, we believe it should also be the most reward-  *ing. For more than forty-five years we've worked closely with our homeowners to bring their fondest dreams to life. And we've come to real- ize that our own success story has a lot to do with our ability to help you build yours. So we'd like to share some of that story with you. It begins in a lush, green corner of the world - with a noble wood.*

THE MAGIC OF CEDAR

*E*very Lindal home has its roots in the rainforests of the Pacific Northwest Coast – the only place on earth where the Western red cedar thrives. The timeless appeal of this fragrant, fine-grained wood is visible in its natural radiance, warm range of colors and velvety finish, all of which make it a daily luxury to live with.

Yet cedar's beauty is more than skin deep; it is also one of Nature's most perfect building materials. Pound for pound, it is as strong as steel, with remarkable insulating properties and a natural resistance to weather, decay, pests and climatic extremes that assures a lifetime of low maintenance and high value.

Lindal's unique system of post and beam construction takes advantage of the wood's strength and glowing good looks. By freeing the walls from bearing any structural load, post and beam opens up all kinds of interior possibilities, including an airy, spacious design that showcases the cedar and lets you arrange interior walls, doors and windows to meet your most personal plans.

Western red cedar: at Lindal, it's the stuff dreams are made of.

Our wondrous, renewable resource: in its native habitat of the Pacific Coast, the mature Western red cedar *(thuja plicata)* reaches heights of 200 feet and diameters of up to eight feet. Botanically classified as an *arborvitae,* or *tree of life,* its visual characteristics include a flared base and thick, fibrous reddish-brown bark. Cedar's high percentage of heartwood and phenol preservatives account for its long life and natural resistance to insects and decay.

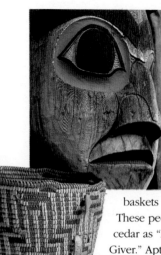

Western red cedar is a wellspring of Native American myths and legends. It held great healing and spiritual significance for the Northwest Coast Indians, who used it in everything from houses and canoes to clothing, baskets and ceremonial objects. These people honorably addressed cedar as "Long Life Maker" and "Life Giver." Aptly enough, the actual native word for the tree translates as "dry underneath."

Under a microscope, a cross-section of cedar reveals the secret of its exceptional insulating ability: nearly 10 million tiny air-filled cells in every cubic inch. Its uniform cell structure gives cedar 12 times the insulating value of stone or concrete. It also contributes to the wood's easy workability and high stability against shrinking and swelling.

Massachusetts

Kiln drying – in our own kilns – adds to the natural advantages of Lindal cedar; it virtually eliminates the moisture that can cause green or air-dried lumber to twist and warp. In the process, it significantly reduces the weight of the wood which, in turn, reduces shipping costs, keeping Lindal cedar homes a competitive value for our customers in even the most faraway places.

Roof Beam

Supporting Post

Floor Beam

KILN-DRIED

Here's the key to the design flexibility of every Lindal cedar home: the time-proven post and beam construction of North American masterbuilders. The weight of the roof is supported by a strong framework of posts and beams, instead of the interior walls on which conventional construction depends. The benefit: your imagination is virtually the only limit to the floor plan possibilities.

Cedar is a natural with Lindal's post and beam construction, which allows soaring open spaces that celebrate the beauty of the wood – in colorations ranging from pale gold and burnished red to deep brown.

As gratifying to work as it is to look at, Western red cedar is easy to cut, easy to nail, and hard to go wrong with. It machines easily with hand or power tools and planes to a rich, lustrous finish. Free of pitch and resin, the wood can be beautifully finished with a wide variety of stains or paints. Left unfinished, it can age to a soft, silver patina.

Quebec

The generous use of cedar detailing radiates natural beauty and quality throughout every Lindal home.

Sir Walter Lindal, the founder of Lindal Cedar Homes and visionary behind the company's reputation for excellence, which has been established over nearly five decades and tens of thousands of custom cedar homes.

Today, all four of Sir Walter Lindal's children (from left to right: Doug, Bonnie, Robert and Marty) are actively involved in the company's daily management; son Robert Lindal has been President and CEO of the company since 1981.

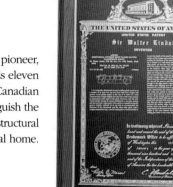

Inventor and industry pioneer, Sir Walter Lindal holds eleven U.S. patents and two Canadian patents that distinguish the building system – and structural integrity – of a Lindal home.

Lindal's international network of local dealers continues to grow along with its leadership in the custom home industry. Locating our sawmill, drying kilns and manufacturing facilities in cedar country allows us to ship all over the globe and remain competitively priced.

FROM OUR FAMILY TO YOURS

*L*indal is the largest manufacturer of custom cedar homes in the world today. Ever since our modest beginnings in Toronto, Canada in 1945, founder Sir Walter Lindal seems to have had an uncanny vision of the company's future and how to get there – even when "getting there" meant moving clear across North America to be close to the source of his treasured building material.

Lindal's move in 1962 to the cedar country of the Pacific Northwest allows the company to serve a worldwide marketplace from one manufacturing base at prices competitive with locally built wood houses anywhere. That makes a difference to our homebuyers.

So does our founding philosophy, which is based on a deep belief that the human spirit is nurtured by highly personal homes, not cookie-cutter houses.

Lindal's size and experience allow us to give our customers the best of both worlds: customization and control. Lindal homes combine the one-of-a-kind style of your own customized plans and the benefits of our design expertise, patented building methods and relentless quality control from forest to building site.

The result is an original that stands the test of time – from a company that does, too.

LINDAL MILESTONES

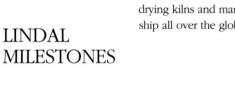
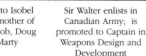

1919	1938	1941	1942	1945	1962	1963
Sir Walter Lindal is born in Saskatchewan, Canada	Early career in the lumber business begins while burning the midnight oil as an architecture student	Marriage to Isobel Rendall, mother of Bonnie, Bob, Doug and Marty	Sir Walter enlists in Canadian Army; is promoted to Captain in Weapons Design and Development	Sir Walter leaves the Army following WWII to start his own home manufacturing company in Toronto, Canada	Sir Walter makes his breakthrough move to the cedar country of Vancouver, British Columbia	Lindal begins to establish dealer network across America and makes its first shipments to Europe

Lindal's move to the Pacific Northwest in 1962 led to the company's dramatic growth during the late sixties and early seventies – and the decison to take the company public in 1971. Going public was the culmination of a long-term dream for Sir Walter Lindal, who promptly put all of the proceeds back into the company to achieve yet another dream: Lindal's own cedar sawmill.

Lindal is the world's only cedar home manufacturer to own its own sawmill, in Surrey, British Columbia. It gives us the control to insist on only small logs, which meet our high criteria for quality and our commitment to using a plentiful, renewable resource. It also gives us – and our homebuyers – major advantages in the pricing and reliable supply of cedar in an industry plagued by seasonal price surges and uncertain availability.

Lindal's network of independent local dealers is the key to staying close to our customers. From the first rough sketch to the last finishing touches of construction and beyond, our 350 home dealers and 125 sunroom dealers provide the local service and support every person needs to build a dream.

Lindal Cedar Homes

Justus Solid Cedar Homes

Lindal Cedar SunRooms

Windows by Lindal

Hardwood Flooring by Lindal

1966 — Lindal headquarters moves to the U.S.; the company's patented A-frame hits the cover of *Popular Mechanics,* bringing national fame to the growing company

1969 — Lindal makes its first shipments to Japan

1971 — Lindal goes public; international headquarters is established in Seattle, Washington

1972 — Lindal builds its own sawmill in Surrey, British Columbia

1981 — Robert Lindal becomes President and CEO; Sir Walter moves up to Chairman of the Board and Director of R&D

1982 — Lindal adds sunrooms to its product line

1983 — Acquistion of Justus Homes adds log home market to the Lindal product line

1986 — Sir Walter receives the first Man of the Year Award from the Home Manufacturers Council of the National Association of Home Builders; hardwood flooring manufacturing facility opens in the Renfrew, Ontario plant

1988 — Lindal acquires Northern Sun Company, which adds the SunCurve to its sunroom line and strengthens its leadership in the cedar sunroom industry

1990 — Sir Walter Lindal celebrates his 71st birthday – and Lindal's success over nearly five decades and tens of thousands of homes

Lindal has its own high standards for the cedar we use. And we have the control it takes to meet them, since we dry, plane and cut all our own lumber at our own sawmill. Quality control begins by inspecting the log booms from which we buy our cedar. Once the best of the boom is at our sawmill, Lindal lumber is produced according to stringent specifications for small logs with tight knots. Our graders and inspectors are tougher than the industry grading rules. And they're on hand to assure the quality that sets Lindal homes apart every step of the way – from logs to lumber to milled building components.

Together, Lindal materials, manufacturing methods and craftsmanship make a difference in the lasting value of your home.

Lindal Cedar Homes Warranty

All Lindal materials are guaranteed to be of the kind and quality specified and free from defects. Lindal guarantees to service and replace any material which is defective or below the quality specified at no cost to you. The guarantee period extends for one full year, providing ample time to inspect all materials as the house is built. However, a physical inventory to check for quantity, and for any freight damage to windows, etc. must be made within ten days of delivery. Lindal warrants that, commencing one year after your home has been completed and passed local inspection for occupancy, we will pay for the repair or replacement of major structural defects occurring within the next ten years. Our salespeople will be happy to provide you with full details on the warranty programs.

Robert W. Lindal, President

Lindal backs its homes with the best warranty in the business – a 10 year guarantee against major structural defects. We want our customers to be as confident about the quality of Lindal homes as we are.

At Lindal, caring about quality has its rewards, and satisfied customers are the greatest reward of all. We share their stories in our *Cedar Living* newsletter for Lindal homeowners.

Cedar Living

Second Annual Photo Contest Winner

LINDAL HOMES: QUALITY FOR LIFE

Lindal's wide world of possibilities begin here, with an introduction to our four major home styles. You'll find more detailed information on each home style in the Specifications chapter, starting on page 130. But this is a good place to get that first overall impression. As you do, keep in mind that you can combine any floor plan and any roof style in this book with any of these four looks. And, because you can modify any Lindal plan to meet your personal dream, the thousands of combinations you can create right from this planbook are just the beginning of the possibilities.

Yet for all of their diversity, Lindal homes have much in common. No matter which home style you choose, you can count on premium Lindal cedar and craftsmanship. High energy efficiency. Low maintenance. Flexible floor plans, thanks to post and beam construction. A solid, heavier home than conventional custom homes. A great value now – and a high resale value later.

Another common feature you can count on: a guarantee that shows just how firmly we stand behind every one of our homes. When you're making one of the biggest investments of your lifetime, that's a nice thing to know.

INTRODUCING OUR FOUR MAJOR HOME STYLES

1

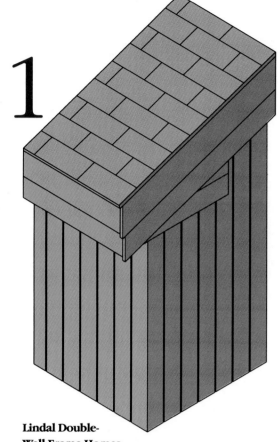

Lindal Double-Wall Frame Homes

This is the original Lindal – a double-wall frame construction with cedar cladding on the outside, drywall on the inside. Lindal homeowners tend to bring at least some premium grade cedar indoors as an accent. But this style lets you enjoy all the advantages of a Lindal cedar home – superior insulation, low maintenance, quality craftsmanship, a cedar exterior – with as much or as little interior cedar as you like. This flexibility of materials and floor plans makes virtually any look possible, from warm and woodsy to cool and contemporary. Maybe that's why Lindal is our most popular style.

2

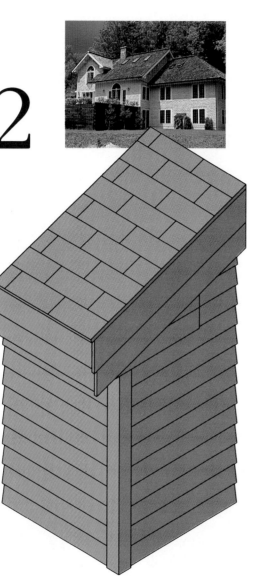

Clapboard Homes

What was once a special option has become a Lindal standard. By popular demand, we offer the traditional clapboard style – with the same superior materials, workmanship and insulation of our original Lindal home. Our clapboard is made of Western red cedar, selected and milled to produce the long, narrow boards that overlap to produce this classic New England look. You may know of clapboard as weatherboard or bevel siding. By any name, it's a well-loved style – available with any Lindal plan, and any modifications, you like.

3

Round Log Homes

Lindal round log homes don't just look warm and cozy. We're proud to say we've captured their rustic appeal without sacrificing our standards for energy efficiency. We set out to overcome the under-insulation and draftiness plaguing most log homes. The result is a natural wonder of uncommon beauty *and* energy efficiency. The airtight fit and super insulation of kiln-dried cedar makes a difference; so does the added insulation between the rounded cedar siding outside and the inside wall.

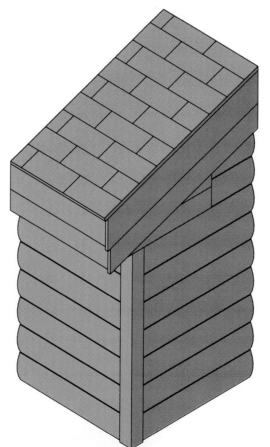

4

Justus Single-Wall Solid Wood Homes

For serious cedar lovers, nothing compares to our single-wall, solid wood Justus. This is a more rustic home, yet it is finely finished like furniture, crafted of four-inch thick cedar timbers that take the place of conventional framing. The kiln-dried timbers lock together for a zero tolerance fit, increasing the energy efficiency provided by the sheer thermal mass of this heavyweight home. Our Justus homeowners tell us how easy it is to build a Justus. And its low maintenance gives you plenty of time to sit back and enjoy its warm, aromatic beauty.

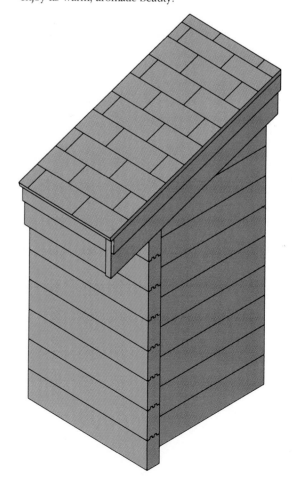

THE POSSIBILITIES ARE ENDLESS

*B*ecause Lindal gives you the flexibility to modify any home plan or create your own one-of-a-kind, there are as many different Lindal homes as there are Lindal homeowners.

It all starts with our unbeatable diversity of designs. You'll find dozens of Lindal home designs with deep historical and regional roots. Others are classic contemporaries – some of today's finest designs for living. Still others are modern originals – truly one-of-a-kind works of art. All of which goes to show there's no end to Lindal style.

Then there's size – just about any size you like. Lindals are dedicated to living well on any scale, from a compact 720 square feet to a sweeping 4,700 square feet, or more. That's because Lindal designs are space-wise, whether you have a little or a lot of room to work with.

Pick a location – any location. In any climate. You can live in a Lindal virtually anywhere in the world – whether it's a tropical island, an alpine mountain, or somewhere in between.

Here's just a sampling from our files of how beautifully Lindal can meet your most personal wants and needs for style, size, climate – and, of course, for the way you live. We think you'll find dramatic proof that your possibilities are as wide as the world.

Cedar
Lindal's Western red cedar exteriors are a wonder of beauty, durability and low maintenance. You can leave this natural beauty unadorned – or add a personal touch to your home's exterior with your own choice of accent materials:

Stone Accents
Local stone emanates strength and lasting appeal.

Brick Accents
New or vintage brick creates a rich traditional look.

Stucco Accents
Stucco can add drama and regional style – perfect for the southwest.

A modern mix: cedar exterior; drywall interior.

Cedar comes inside – a little or a lot.

4/12: a low pitch roof line.

8/12: a moderate pitch.

Any roof works on any plan.

Any roof can top a one-of-a-kind.

Passive solar captures the sun's energy.

Sunrooms bring the outdoors in.

Solid cedar inside and out.

Lindal exterior with brick.

Lindal exterior with stone.

Lindal exterior with stucco.

12/12 roof - for cathedral ceilings.

Traditional gambrel roof.

Classic hip roof.

Combine roof styles.

Cool in hot, arid climates.

Resists moisture and decay in the tropics.

Pole homes for sloping sites and flood zones.

Let it snow! It's warm and snug inside.

Sunrooms do wonders for business.

Presidential Prow home in Japan.

Commercial success in Japan.

Lindal means business.

From moist tropical regions and hot, arid deserts to cold, snowy heights, a Lindal is at home just about anywhere on earth. Designed to take full advantage of cedar's insulating properties, Lindal homes stay warm in cold weather, refreshingly cool in the heat. The natural preservative oils in Lindal cedar also resist the ravages of climatic extremes, moisture and decay.

*Y*ou can have just as much – or as little – involvement as you desire in planning your Lindal home.

PLANNING YOUR DREAM

1 If you're looking for the convenience of a stock plan, you've come to the right place: Lindal offers the widest selection of cedar home styles and floor plans in the world.

2 If you want a proven home design that you can modify according to your personal wants and needs, you'll feel right at home with Lindal; after all, this is the way most of our customers choose to work with us.

3 Every year, more and more people come to Lindal to create their one-of-a-kind homes. Your ideas and our quality materials and craftsmanship make an unbeatable team. If you're working with architects, we're happy to work with them, too.

*P*lanning the home you've always dreamed of is probably one of the most exciting things you'll ever do. And the rewards last a lifetime. But it's not always easy. That's why Lindal has a worldwide network of local dealers – experts who can help you every step of the way in bringing your Lindal to life.

For Jeff and Tiny Rubenstein of central New York, working with their Lindal dealer was "like having our own building consultant - at no extra charge.

Ours was a custom designed house. Our dealer, Jim Johnston, was a great help, from initial design right through construction. He flew up to visit the building site, sat down with us, and drew up the first sketch (on notebook paper) of the floor plan, while we prompted him with our various ideas and wish list. What a thrill it was when the first set of detailed plans came back from Seattle!

We acted as the general contractor, and Jim was particularly helpful during the construction process... Of course, he promised this sort of service initially, but it turned out to be more than just a sales pitch – he followed through admirably."

Ontario

WISH LIST!

2 Car Garage
4 Bedrooms
Big Kitchen!

Den
High-Lofted Ceilings
Walk-in Closet
Marble Fireplace

Sundeck
Formal Dining Room

A "wish list" is an exciting – and important – first step in planning your Lindal home. Getting your dreams down on paper helps to start the creative juices flowing and gives you a valuable point of reference as you move through the planning process with your local Lindal dealer. A friendly word of advice: don't try to "edit" your dreams at this early stage. You'll be surprised just how many of them can come true as you plan your Lindal home.

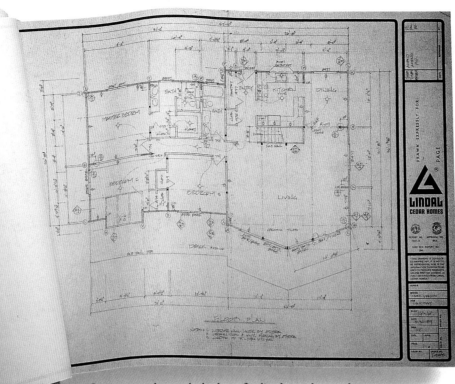

Once you and your dealer have finalized your home plans, Lindal designers transform them into architectural blueprints – complete with every specification needed to begin construction. Final working blueprints are keyed to the numbering system on your materials list and the building components themselves. This identification system streamlines construction by making it quick and easy to get your hands on the building parts you need – when you need them.

Our homebuyers are happy to learn that all Lindal building materials can arrive in one convenient shipment – or two, upon request. Compare this with the unreliable supplies and schedules so common with custom builders today – and rest assured you can count on Lindal.

The rewards of all your planning really come home when you move into your Lindal and begin to share it with the people you love.

Lindal's Engineering Department is a close-working team that combines time-honored draftsmanship with state-of-the-art Computer Aided Design (CAD) systems to generate the blueprints of your dream – whether it's a custom modification to a Lindal plan or your own one-of-a-kind.

CHAPTER TWO

ONE-OF-A KINDS

Every Lindal home is a custom home. While some are personal variations on a Lindal floor plan, others are totally unique. So if your notion of a dream cedar home is one *that you or your architect designs, you're in good company, as you can see from the one-of-a-kind homes in this chapter. What's more, you'll be happy to know that you don't pay a premium for the luxury of your own original masterpiece. A modest design fee, based on the square footage of your home, is all it takes to translate your ideas into working plans and final blueprints.*

ROMANCING THE RANCH

❧

*N*orth of Santa Barbara, in the Santa Ynez country, this rambling ranch sits on a hill in classic California ranchland, where rolling brown hills and wild oak provide a picturesque setting for raising horses and llamas. The owners designed their ranch home to suit their casual lifestyle elegantly – and to take in the views all around them.

This is our cover home. Another exterior photo is shown on pages 16-17.

The master bedroom suite occupies its own wing of the house – with a sumptuous bath and adjacent study.

An antique chest, floral wallpaper and a bouquet of roses express personal style in the master bath.

ONE-OF-A-KIND #1

- 3 Bedrooms
- 3 full Bathrooms
- 2 half Bathrooms
- 4,174 sq. ft.
- Overall Size: 111' x 60'
- Master bedroom on first floor.

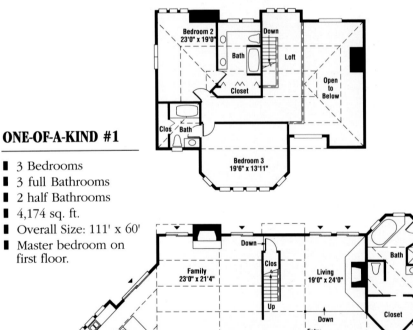

Every room in this ranch home looks out onto views of the Santa Ynez ranchlands below. ➤

DESIGN FOR ENTERTAINING

🌷

*J*ust minutes from the historic town of Wood-stock and the challenging Killington ski slopes, this clapboard classic is at home in its wooded setting. Bob and Judy Newsome designed and built their 3,178-square-foot home for two — with the pleasures of enter-taining friends and family well in mind. The open kitchen and sunroom area is clearly the heart of this home — and the most popular gathering place for guests.

The open kitchen has a relaxed feeling that reflects the owner's love of casual, comfortable get-togethers.

A formal mood prevails in the dining room, with its black lacquered French doors and bleached hardwood flooring.

ONE-OF-A-KIND #2

- 3+ Bedrooms
- 3 Bathrooms
- 3,178 sq.ft.
- Overall Size: 74' x 51'
- Master bedroom on second floor.

Opposite the entry and just off the dining room, the living room has a high cathedral ceiling and a stone fireplace.

*B*usy careers don't slow down Bob and Judy Newsome: Both are expert skiers; Bob pilots his own helicopter; and they love to entertain family and friends in the home they designed together.

A five-bay sunroom and a cantilevered balcony off the master bedroom bring the facade of this clapboard home up to date. ➤

PARADISE FOUND

❦

Dear Mr. Lindal:

It had been our dream to construct our own custom home in Hawaii for some time. But with little experience in building, we were very apprehensive. After meet-*ing with your dealer, Ted Cormier, and having him show us the Lindal method of post and beam construction, and the quality of a cedar home, along with the versatility to truly design a custom home to fit our lifestyle, we took a deep breath and committed ourselves to what has turned out to be a deeply satisfying experience. Your company's services and product were all that were promised, and more!!*

Ted Cormier was especially help-ful in every phase of the project. He was there with help, advice, encour-agement, and even loaned us his tools at times. He was there long after his contractual obligation was completed. His experi-ence in design and con-struction, along with his enthusiasm, helped us realize our dream, and we will be forever grateful.

Thank you all; we feel we were very fortunate to have chosen Lindal Cedar Homes!!!

Sincerely,

Al and Naomi Weakley

Kailua Kona, Hawaii

Double doors provide a wide-open welcome to this 2,831-square-foot home overlooking Kailua-Kona and the blue Pacific.

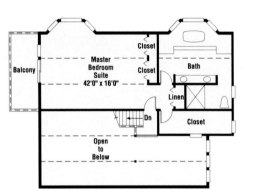

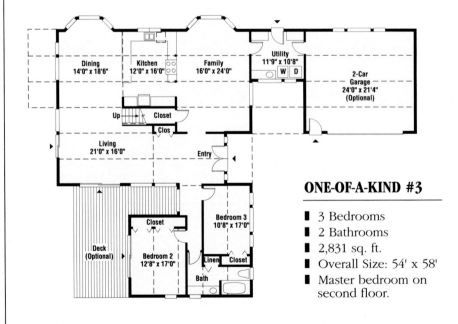

ONE-OF-A-KIND #3

- 3 Bedrooms
- 2 Bathrooms
- 2,831 sq. ft.
- Overall Size: 54' x 58'
- Master bedroom on second floor.

Varying roof lines add to the visual appeal of this home, with its driftwood siding. ➤

NEW ENGLAND LIGHT

❦

When Jim and Linda Hargrove decided to build a home, Jim spent a year analyzing the options. He looked at floorplans, interviewed architects, and researched manufactured housing companies. In the process, he met local Lindal dealer Dom Vingiano – and found the working relationship he was looking for.

Collaborating with Vingiano, Jim said, gave him the opportunity to be his own "dream architect." Vingiano adapted Jim's creative ideas into cost-effective reality.

"Dom was unusually helpful from design through construction. He and Lindal provided a complete service for the person who wasn't sure exactly what he wanted to do."

The Hargroves' experience with Vingiano – and the end results – were such a positive testimonial that their neighbor has since built a Lindal home.

But perhaps the strongest testimonial to the home is Jim's reluctance to leave it for their vacation house in Maine. "We have a beautiful summer home on the water there, but it's such a pleasant environment here, so open and spacious, we'd rather stay home."

In its New England setting ablaze with autumn color, the Hargrove home is a masterpiece of cedar and light.

ONE-OF-A-KIND #4

- 3+ Bedrooms
- 2 1/2 Bathrooms
- 2,358 sq. ft.
- Overall Size: 53'x 43'
- Master bedroom on second floor.

A six-bay sunroom on the second level is the focal point of this bleached wood home in the desert; grids on the sunroom glass are a personal accent.

DESERT OASIS

❦

*T*he owners of this Arizona home were determined to make their little piece of desert bloom. And it does. Gardens are terraced down the slope of the property. A grape arbor flourishes to the side of the home. Rugged stone blocks form easygoing pathways and steps. And everywhere there's a profusion of floral color, created by in-ground plantings and terra cotta pots.

ONE-OF-A-KIND #5

- 3+ Bedrooms
- 2 1/2 Bathrooms
- 3,079 sq. ft.
- Overall Size: 50' x 48'
- Master bedroom on second floor.

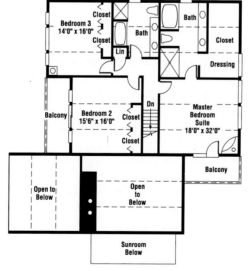

The masonry fireplace provides a cheery blaze on cool desert nights – and the sunroom opens up to a view of starlit skies.

New Jersey

A Stone Cabin Grows Up

❦

*D*ear Mr. Lindal:

Our Lindal was an addition to an existing home that I purchased from my great uncle's estate. The original house was a 40-year-old stone ranch house with a dormer in the back. I believe the original house was about 1,200 square feet. It is now 4,740 square feet. So it ended up being quite an addition.

Wayne and Anita South of Cedar Classics contacted us when they heard we were working on an addition. Your catalog, along with Wayne and Anita's endorsement, went a long way to convince us that a Lindal addition was a perfect match to our stone house. They certainly were correct.

It is now a home with plenty of space, yet everyone feels comfortable. They say "it's homey." That's a nice comment, one you do not often get for a large home. So what do we like

best? I guess the answer is the warmth.

Is it energy efficient? Truthfully, it seems to cost me less money to heat than the 1,800-square-foot house I previously owned.

Did we design our own home? Absolutely. With Wayne and the architect's help; we were able to build what I call a "man's dream."

I was very worried when I started this project that I would mess up what was already a beautiful stone house. Instead, we created the "dream house" my wife and I had always wanted.

I would recommend both Lindal Cedar Homes and Cedar Classics to anyone who asks, and many have!

Scott and Henrietta Welch,

N.W. New Jersey

■ 4 Bedrooms
■ 2 1/2 Bathrooms
■ 4,740 sq. ft.
■ Overall Size:
   79' x 66'
■ Master bedroom
   on second floor.

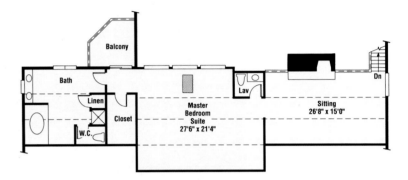

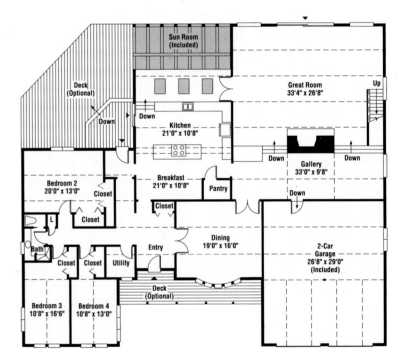

Local stone adorns the home, with cedar clapboard on the other three sides. ➤

A MODERN LOG HOME

❦

"*I was drawn to Lindal because of the attractiveness of the particular type of solid cedar logs you produce. I wanted a log home for reasons of lifestyle and efficiency of construction.*" That's how the owners describe the impulse behind their Justus home. Their one-of-a-kind combines the best of classic ranch style with contemporary flair, and shows how modern a Justus home can look.

The home embraces a terrace with a spectacular Puget Sound view.

ONE-OF-A-KIND #7

- 2 Bedrooms
- 2 Bathrooms
- 2,463 sq. ft.
- Overall Size: 98' X 50'

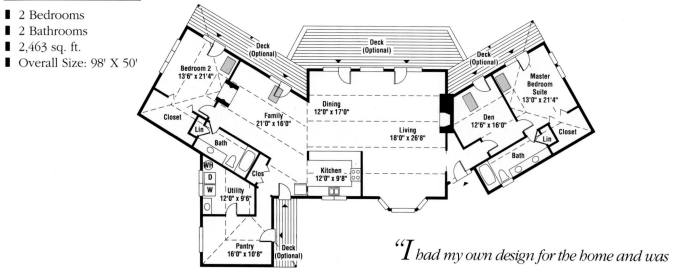

Bleached cedar timbers gives this Justus a clean, contemporary look that sets it apart from most log homes.

"I had my own design for the home and was pleased to find that Lindal could prepare drawings and specs adaptable to my plan. The help and cooperation of your local dealer was greatly appreciated."

The very structure of the roof beams is attractive as well as strong. ➤

LINDALS FOR BUSINESS AND COMMUNITY

❦

*T*hese days, many professional offices, small businesses and community buildings are locating in residential areas. The more sensitive of these new neighbors strive for new construction that's in keeping with the residential character of the community. And few buildings do this better than a Lindal, which has its roots in residential design but adapts readily from custom residences to hardworking solutions for businesses and communities.

Naturally, Lindal commercial buildings can also be built in commercial locations. In fact, our cedar-framed sunrooms are popular architectural elements of many restaurants, fast food drive-ins, medical clinics and offices.

So bring us your ideas. We welcome the opportunity to make the most of them for your business or community.

ONE-OF-A-KIND #8

- United Covenant Church of Wilton
- 3,339 sq. ft.
- Overall Size: 98' X 53'

*"W*e would like to express to you our appreciation for the services provided by Lindal Cedar Homes and your local dealer. An innovative church building design was possible, incorporating the post and beam concept. It was a pleasure to work in such close harmony with you."

Pastor Gordon Miller, Wilton, CT

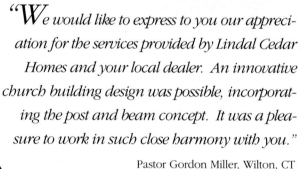

When the time came to expand, the Off-the-Cuff Restaurant in Wellesley, Massachusetts took full advantage of the opportunity to add a striking architectural drawing card. They found their answer in this two-tiered SunCurve from Lindal. The cedar-laminated arches curve in a double cascade to the ground.

New Jersey

Steve Savoia was happy with his Lindal home, (below) and more than satisfied with his dealer, Mona Kapper of Cedar Designed Homes. So together, they designed and built this two-story office building (above) on busy Route 9 in Old Bridge, to accommodate Steve's growing business.

New Jersey

Massachusetts

CHAPTER THREE

CLASSIC HOMES

Lindal's collection of classics represents some of the most appealing styles in residential design today. For all their diversity, these homes share a casual elegance that  *never goes out of style. Many feature multiple levels, long rooflines and flexible interiors that welcome the use of vaulted ceilings, window walls and sunrooms. And there's a classic style for just about any site, from spacious suburban properties to narrow city lots. Their timeless architectural character makes Lindal classics at home in fine residential neighborhoods anywhere.*

DESIGNER

These are modern classics, dramatized by a two-story sunroom which serves as a stunning formal entry. The sunroom is also a cherished living space that lends an airy openness to the entire home. All three Designer floorplans feature a cathedral ceiling in the living room. In two plans, the master bedroom is situated upstairs; the third locates it on the main floor.

A Lindal sunroom brings more than sunshine into your life; it's a magical place to be on a starry, moonlit night.

Washington

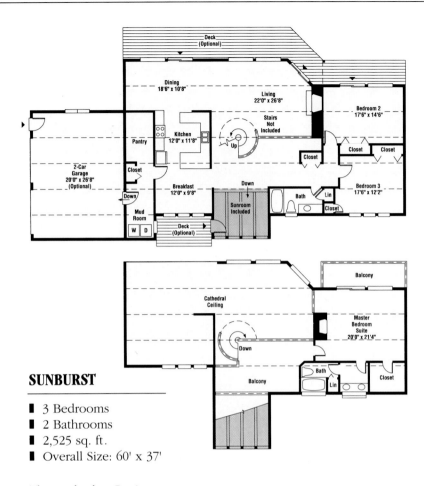

SUNBURST

- 3 Bedrooms
- 2 Bathrooms
- 2,525 sq. ft.
- Overall Size: 60' x 37'

This and other Designer plans are shown in larger scale on pages 153-155.

Washington State may be famous for its rain, but it's been known to have snowy winters, too. The Sunburst shown here and on the previous two pages is located on Vashon Island with an easy commute to Seattle. Yet Jim and Shannon Foster have plenty of room to stable the horses they love, right on their own property.

This Sunburst glows with the warmth of cedar. You can bring the fine-grained beauty of cedar wall and ceiling liner into any room of any Lindal home.

These owners opted for a fresh, bright look, using cedar trim in contrast to the sleekness of white drywall.

"We had a design that we thought was really special, sort of our dream home, and we went to Lindal. They were able to take many of the special things we wanted and incorporate many of the really nice features of the Lindal homes, such as the garden windows and the sunroom, and also the skylight." Shannon and Jim Foster, Vashon Island, WA

High clerestory windows bathe the dining room with natural light. The cedar ceiling in the kitchen radiates a warm welcome to good friends and good food.

The open-baluster balcony leads to the master bedroom upstairs – and adds to the sense of spaciousness in the living room below. Maple hardwood floors wear well – functionally and aesthetically.

"We love the feeling of spaciousness with the open concept and the room it provides for free-flowing movement."

Dean Davey, Port Perry, ON

A cathedral ceiling and large picture windows highlight the living room and make the most of its pastel color palette.

Clear cedar mullions frame a favorite view. The home-owners get a lot of use out of this sunroom entry as a casual, comfortable sitting room.

A mix of roof lines and wide expanses of glass make the back of this Designer home as attractive as the front.

This customized Designer, reminiscent of the Sunglow, features a five-bay sunroom. ➤

CITY

I f you have a city lot with a view, consider this elegant solution; our City series was designed especially for narrow urban sites. The plans present a private facade to the street, while the interior opens up to a two-story sunroom in the living room – all in all, an uplifting perspective on city living. All three plans locate the main living areas on the first floor, with most bedrooms on the upper level.

Our City homes make the most of interior views, too. This open balcony on the second floor overlooks the sitting area in front of the fireplace.

BOSTON

- 3 Bedrooms
- 2 1/2 Bathrooms
- 1,624 sq. ft.
- Overall Size: 35' x 31'

This and other City plans are shown in larger scale on pages 156-158.

In the dining room, patio doors open onto a much-used second-story deck.

This young family enjoys an outstanding view of downtown from their City home high on a hillside.

Its garage shelters this narrow City house from the street. High windows serve double duty in this design, allowing lots of interior light while insuring privacy.

Urban elegance: the owners painted all interior wood to match the sophisticated spirit of their City home's decor. ➤

"Of course, there's always a sense of pleasure in hearing a gasp of surprise the first time a visitor steps through the front door and experiences our home. But what's most precious are those moments of relaxation, just sitting in the living room, for example, enjoying the beauty of nature outside and the heavens above. It's a feeling of peacefulness."

Philip and Barbara Bast, Waterloo, ON.

Generous-sized decks can expand living out-of-doors.

With a kitchen like this, cooking up new recipes is a pleasure. Mini blinds in the sunroom facing the street can be adjusted to control light and privacy.

A sunroom rising to the sky is the focus—and the joy – of this living space. ➤

EXECUTIVE

Here are the landmark homes of their neighborhoods: our Executive series. Their street appeal includes stepped-down roofs, clerestory windows and the symmetry of the front entry. Inside, the central foyer opens up to a two-story cathedral ceiling, introducing a roominess that is carried through in every interior space. Executive homes range in size from 1,760 to 3,394 square feet.

This dining room is enriched by the warm, rich colors of its cedar ceiling liner and cedar framed windows.

A screened porch makes outdoor eating and relaxing a summertime pleasure you can count on, uninterrupted by rain or insects.

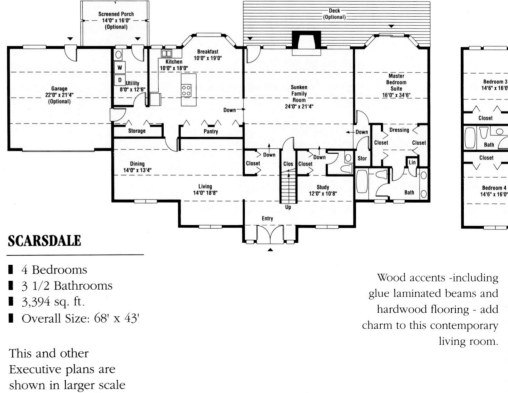

SCARSDALE

- 4 Bedrooms
- 3 1/2 Bathrooms
- 3,394 sq. ft.
- Overall Size: 68' x 43'

This and other Executive plans are shown in larger scale on pages 159-161.

Wood accents -including glue laminated beams and hardwood flooring - add charm to this contemporary living room.

This home, built near Cincinnati in the 1970's, was the inspiration for our Executive Series. ➤

Our cedar and glass sun-
rooms blend beautifully with
our cedar homes. For more
information on sunrooms
turn to pages 128-129.

*"We love our sunroom and the open-
ness of our Lindal. It gives a feeling of
out-of-doors in any season."*

M.O., NY

Create your own comfort zone: this New York family customized their Georgetown by adding a sunroom and family eating area off the kitchen.

Massachusetts

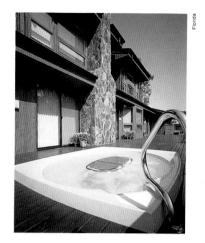

Florida

A hot tub with jacuzzi makes deck living even more appealing; if there's a hot tub in your future, plan ahead for structural strength.

"We chose the Polar insulation package, and our annual heating costs have been less than we paid in our previous home, although our Lindal home has 50 percent more floor space."

Mr. and Mrs. John W. Henderson, Jr., Hyde Park, NY

These ingenious home-owners created a special kids' room upstairs, with comfy platforms, bunk-beds and plenty of nooks and crannies for playing.

Florida

New York

This Executive home happens to be in Florida. But the series is designed to make an impressive con-tribution to any upscale neighborhood – anywhere in the world.

In our Georgetown and Scarsdale plans, the master bedroom has a wing of its own on the main floor; the rest of the bedrooms are upstairs.

CONTINENTAL

*T*he pleasing and practical character of this series comes from having zoned living activities on different levels. Each of the three floor plans achieves this in its own unique way. But in all cases, a high ceiling enhances the spacious feeling of the living and family area on the main level, while the master suite and other bedrooms take advantage of the high peaked ceilings and open beams on the top level. Continental homes range in size from 1,615 to 2,273 square feet.

The back of the Bavaria opens to the pool. Lustrous cedar decking is a natural complement to the pool and the home.

Livable design on every level: the lower living area of the Bavaria is separated from the main living area by stairs, yet the two areas are open to each other.

BAVARIA

- 2+ Bedrooms
- 2 3/4 Bathrooms
- 2,166 sq. ft.
- Overall Size: 43' x 44'

This and other Continental plans are shown in larger scale on pages 162-164.

There are lots of variations on the Continental theme. This is the Burgundy.

Two shed projections on the back of the Bavaria plan distinguish it from the other two Continentals. ➢

INTERNATIONAL

*I*n this ever-popular series, two wings join at right angles to form an "L" with bedrooms in one wing, living areas in the other.   Many Internationals are built as one-story ranch homes or ramblers. Others are built on sloped lots, so the plans provide for stairs, which make it easy to add a walkout daylight basement below.

A walkout daylight basement doubles square footage for a relatively modest increase in construction costs.

A swimming pool, patio and lavish deck add to the enjoyment of this home.

TOKYO

- 3 Bedrooms
- 2 Bathrooms
- 2,627 sq. ft.
- Overall Size: 66' x 56'

This and another International plan are shown in larger scale on pages 165-166.

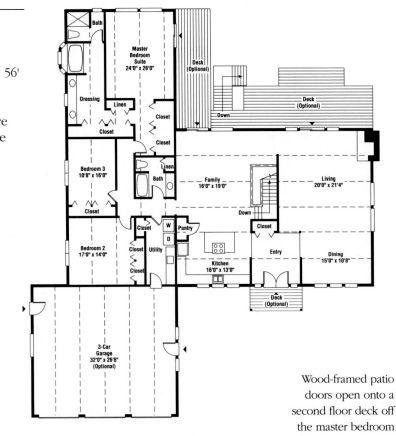

Wood-framed patio doors open onto a second floor deck off the master bedroom.

The glow of cedar at dusk is part of the daily beauty of this home at Solvang in the Santa Ynez ranchlands of California. ➤

SIGNATURE

🐎

*T*he multiple roof pitches and covered entries of these homes give them an attractive street presence. But they're designed for privacy, too. Inside, Signature homes open to the back, which makes them ideal for view sites. The master suite is located on the main floor of one plan, the second floor of the other.

Hamlin's Pond, a saltwater inlet on Cape Cod, is the serene setting for this home.

Massachusetts

To make the most of an outstanding view, sunwalls top a bank of sliding glass doors, which lead from the living and dining areas to outdoor decks.

*"T*he plans on paper cannot do justice to the beauty, structure and comments from our friends on our Lindal. It's a dream come true!!"

Guy and Ann Brigida, Mashpee, MA

HALLMARK

- 3 Bedrooms
- 2 1/2 Bathrooms
- 2,325 sq. ft.
- Overall Size: 50' x 35'

This and another Signature plan are shown in larger scale on pages 167-168.

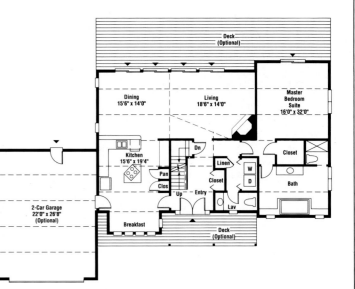

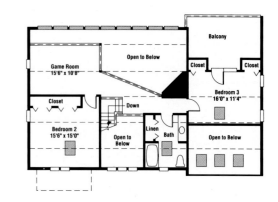

More photographs of the Hallmark are shown on pages 130-131 and 226-227. ➤

EMBASSY

🐴

With their high shed roofs and absence of roof overhangs, these homes evoke historic salt box design. Their satisfying symmetry is enhanced by a central entry with double doors, skylights overhead and high clerestory windows. Inside, the formal living and dining areas are on the street side; casual family living areas are to the back of the home. Embassy plans range in size from 2,954 to 4,420 square feet.

The master bedrom suite has a wing all its own on the main level; other bedrooms are located upstairs.

DIPLOMAT

- 4 Bedrooms
- 3 1/2 Bathrooms
- 4,420 sq. ft.
- Overall Size: 72' x 51'

This and other Embassy plans are shown in larger scale on pages 169-170.

"We love the cedar siding. Everyone loves the kitchen, and the open spaciousness of the house is fabulous! The company has been very pleasant and Jim Johnston was the best dealer. He always had time for us."

Peter LoPiccolo, Armonk, NY

The cozy family room, with its big stone fireplace, is just footsteps away from both the kitchen and deck.

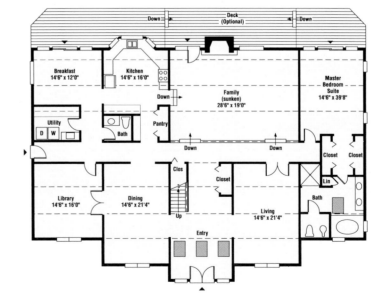

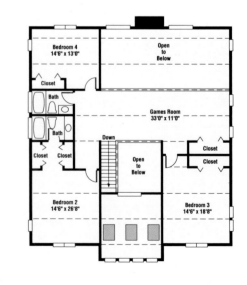

This home on its large wooded lot is only a half hour drive from Manhattan. ➤

TOWN & COUNTRY

*I*n city and country alike, property prices are increasing and small, narrow lots are often the norm. Like our City series, Town & Country homes are designed to fit a lot of living space into a narrow site. The street side is designed to protect privacy; the other side has an all-glass prow front that opens the view to the back of the property.

The street-side garage provides privacy; its roof extends on the right to create a covered walkway.

From the street, no one would guess that this home opens up with a spectacular all-glass prow front – and a sunroom.

California

FAIRLANE

- 4+ Bedrooms
- 3 Bathrooms
- 3,388 sq. ft.
- Overall Size: 51' x 37'

This and another Town & Country plan are shown in larger scale on pages 171-172.

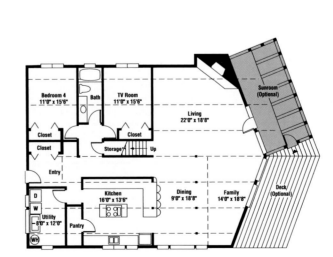

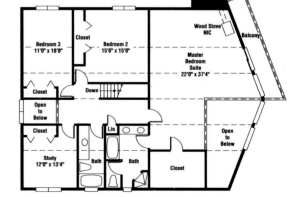

The living room looks out onto a golf course near prime ski slopes in the Sierra Nevada Mountains. ➤

CENTENNIAL

The central entry is the focal point of our Centennial homes – inside and out. It rises to a lofty two-story height and provides convenient access to all three levels. Bedrooms are located on the top level. Of course, with a basement provided for in the plans, the possibilities for expansion are boundless. The Centennial plan is 3,427 square feet and the other is 2,172 square feet.

Massachusetts

New York

Here's just one inspiration for finishing off a basement in style. How about an aromatic cedar-lined sauna?

CENTURY

- 3 Bedrooms
- 2 1/2 Bathrooms
- 3,427 sq. ft.
- Overall Size: 61' x 37'

This and another Centennial plan are shown in larger scale on pages 173-174.

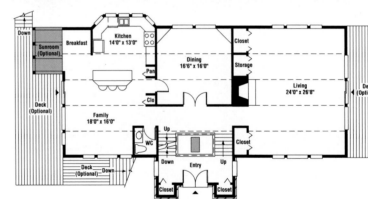

"I would recommend Lindal to my friends, or to anyone who admires my home. It is our investment in our life's dream house."

A. and M.G., NY

Connecticut

Raised panel bifolds, in clear cedar, are an option that makes a special children's bedroom even more so.

Brick is a handsome accent to this cedar home in New York State. ➤

MASTERPIECE

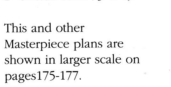

We've recaptured the romance of pavilion archi-tecture – and created these Masterpieces in the process. The multi-sided pavilion, beautifully integrat-ed into the design, is the hallmark of this series. In two of the plans, the pavilion is dedicated to family living areas. In the third, the master bedroom has the place of honor. All three plans locate additional bedrooms on the second floor.

With both an eating bar and breakfast table, the kitchen is designed to meet the needs of a growing family.

"*We like the design, the flexible interior solutions offered by the Lindal concept, and the quality of the wood products offered.*"

Tove & Terje Gilje,
Aurora, Ontario

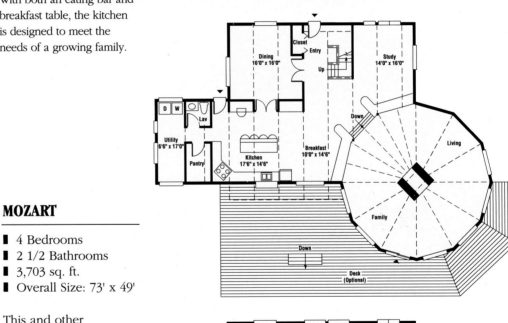

MOZART

- 4 Bedrooms
- 2 1/2 Bathrooms
- 3,703 sq. ft.
- Overall Size: 73' x 49'

This and other Masterpiece plans are shown in larger scale on pages175-177.

The ceiling structure in the pavilion is a thing of beauty itself. Cedar ceiling liner and hardwood flooring add to the allure of this space. The central fireplace is open on both sides.

Expanses of cedar decking extend living out-of-doors, and lead to the swimming pool. ➤

CHAPTER FOUR

CONTEMPORARY HOMES

Allow us to introduce our most dramatic statements: Lindal's contemporary homes are the height of bold design. Imagine soaring roof lines. Panoramic prow fronts. Walls of windows. Cathedral ceilings that create a sweeping interior spa- 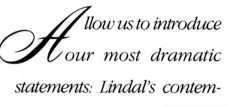 *ciousness. Yet their stunning appearance is just the beginning of their beauty; Lindal contemporaries are intelligent designs for living. Their expansive sense of freedom is much more than a look; it's a way of life that contemporary home-owners love to come home to.*

PROW STAR

These are our best sellers - and with their glass-fronted prows offering sweeping, two-story views, it's easy to see why. Our Prow Stars' roomy designs include lower-pitched wings on one or both sides. You'll find a wide range of plans for these homes in our design library, ranging in size from 1,387 to 2,021 square feet. But, as our Prow Star homeowners have proven, the possibilities are endless.

New York

"Our home seems to be 'famous' in the town, as it is so attractive and different. Yet it blends in beautifully. We are very proud."

Peggy and Bob Crespo, Thornwood, NY

New Jersey

Diagonally-placed cedar creates a striking accent on the stairway wall leading to the upstairs loft.

SEA VISTA

- 3 Bedrooms
- 2 Bathrooms
- 1,834 sq. ft.
- Overall Size: 60' x 36'

This and other Prow Star plans are shown in larger scale on pages 178-184.

This and other Prow Star plans are shown in larger scale on pages 178-184.

* The term "Star" indicates that the homes in this series have wings.

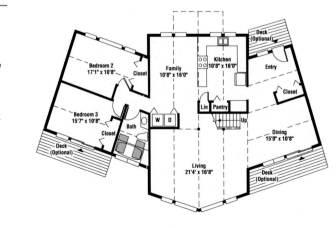

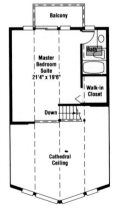

Building on a walkout daylight basement can double your living space economically. ➤

California

Designed for privacy and convenience, the master bedroom leads to a compartmentalized bath with separate areas for a walk-in closet, vanity and lavatory.

California

There are no artificial boundaries between the dining room and this kitchen, with its island eating bar. The red and green color scheme is carried out throughout the home.

Finely-finished solid cedar timbers and an inviting window seat help make this home on the Truckee River cozy and comfortable all winter long.

California

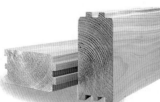

"I have personally enjoyed this home from an aesthetic standpoint mostly. However, after living in a cedar log home, I have learned to expect the comfortable, even temperature in summer or winter."

B. K. L., Paso Robles, CA

This Prow Star reflects the beauty of its location on Big Rideau Lake, south of Ottawa. ➤

New York

New York

The master bedroom looks out onto the living room below – and the view beyond.

"Our Lindal home has created for us a feeling of being at one with nature through its openness and large expanses of beautiful windows. It's a joy to live in."

Sally Weilberg, Westerly, RI

"We have lived in our Lindal home for over three years now, and are very happy we chose Lindal. Every winter here is severe, with many days and nights of bitter cold and extreme winds. With our Polar insulation, the house stays comfortable, and our heating costs are very reasonable."
Jeff and Tiny Rubenstein, Central New York

Connecticut

A skylight and cedar framed windows add light and ventilation to this sleek, green-tiled bathroom.

Rhode Island

The use of local stone from Esterel on Lac Croche (Crooked Lake) is a natural with Western red cedar, which radiates a spectrum of colors from pale golds and burnished reds to rich browns.

"I would recommend Lindal for its quality, its design flexibility, its respect of our agreement, especially the delivery date, and, finally, its customer after-sale service."

Jean-Rene Fournelle, Ste-Marguerite, PQ.

The street side of this customized Prow Star has a recessed formal entry faced with stone.

This small, three-bay sunroom is an architectural element that adds beauty as well as light and ventilation to the dining room.

New Jersey

A skylight over the four-poster bed encourages stargazing and sweet dreams. Here the finely-planed solid cedar timbers have been bleached for a lighter look.

LAKE VISTA

- 2 Bedrooms
- 2 Bathrooms
- 1,976 sq. ft.
- Overall Size: 47' x 37'

This and other Prow Star plans are shown in larger scale on pages 178-184.

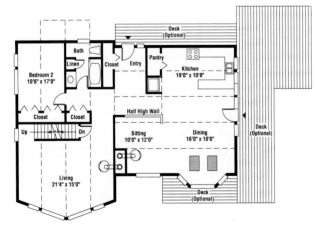

The owners finessed this small bathroom off the master bedroom into a real charmer.

New Jersey

"*Our very first home was a charming Justus cedar log home. We loved the beauty and warm richness of the wood, and we were quite impressed with the energy efficiency of our home. When it came time for us to move, there was no question but that we would design and build our own Lindal/Justus home.*"

Gail Denemark and Gary Ksander, Delaware Valley, NJ

Notice the top-knot over the lower-pitched wing on the right? That's the master bath.

New Jersey

In this bright dining and kitchen area, country decor works hand in hand with the solid cedar timbers and hardwood flooring. ➤

PRESIDENTIAL PROW STAR

Here we've built on inspiration, starting with the same striking prow front as our Prow Star series — and taking the rest of the home to new heights with a two-story wing that adds economical floor space. The roof line of the single-story garage is also extended to create the effect of a covered veranda across the front facade of the two-story wing.

New Jersey

Guests are sheltered from the elements by the covered entry of this "Lincoln" home in Monmouth County.

LINCOLN

- 3 Bedrooms
- 2 1/2 Bathrooms
- 2,047 sq. ft.
- Overall Size: 46' x 38'

This and another Presidential Prow Star plan are shown in larger scale on pages 185-186.

* The term "Star" indicates that the homes in this series have wings.

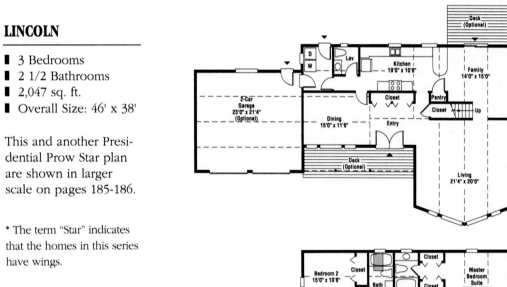

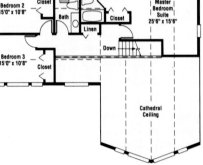

Ontario

The finely-grained cedar ceiling adds a warm ambience to this dining room.

The owners of this customized home in the Albion Hills chose to locate their main entry beside the garage and rearranged the interior spaces to their own requirements. ➤

VARIATIONS

Inspired by our Prow Star and Presidential Prow Star homes, two Lindal homeowners have created their own personal variations on our theme. The D'Argenio home in Connecticut is shown on this page and opposite is the DeBerry home in Massachusetts. They invite you to study their photos as inspiration for your own custom Prow home.

Fine detailing in the woodwork adds to the exquisite decor of this home.

"It was a cold day in February when I first saw the land where we built our Lindal cedar home. It is three acres facing a 20-acre pond. The day I looked at the land there were 2,000 Canadian geese swimming in the pond. I bought the land that day... With the help of our dealer, Art Hawkins, my wife and I designed the perfect house for the perfect location – a Presidential Prow overlooking the entire pond.. Thanks, Lindal, for a wonderful house!"

Raymond and Barbara D'Argenio, CT

Room with a view: the prow windows in the living room bring the outdoors in.

Natural companions: the quarry tile of the dining room floor is another material that's harmonious with cedar.

This home has it all: a large one-story wing with a 10-bay sunroom, and a two-story wing off the prow.

Wood framed sliding glass doors lead to an expansive deck; the brick fireplace complements the Western red cedar.

"Our dealer was very knowledgeable and helpful in the building of our home. He followed through until our home was completed. We are very happy with the service we received during the ordering, building and finishing of our home."

Larry and Laura DeBerry, Hopkinton, MA

A sunroom and hardwood floors make the front entry gleam.

The dining room has a privileged position in this home, opening to both the sunroom and living room.

CHALET STAR

The Chalet Stars are perennial favorites, with their cathedral ceilings and window walls that take in the view. These homes have wings on one or both sides, which encourages zoning of activities. Spacious master bedroom suites are located on the upper level. Chalet Stars range in size from 1,466 to 2,255 square feet.

The whiteness of drywall provides a clean contrast to the richness of cedar trim.

A massive fireplace built of local stone is the centerpiece of this Chalet Star. It's a signature look that's been replicated in countless Lindal homes.

CAPRICORN

- 2 Bedrooms
- 2 Bathrooms
- 1,483 sq. ft.
- Overall Size: 40' x 31'

This and other Chalet Star plans are shown in larger scale on pages 187-192.

* The term "Star" indicates that the homes in this series have wings.

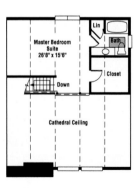

This customized Chalet Star features a five-bay sunroom off the dining and kitchen area for relaxing and entertaining. ➤

The Zodiac plan features two chalets intersecting at right angles, giving the same roof pitch to both wings.

Washington

"We are very proud to own our Lindal home because it is so unique and personal."

Karen and Timothy Hunt, Wappinger Falls, NY

Solid cedar timbers, smooth to the touch, radiate a natural glow in this unique bathroom.

New York

Two wings accommodate a roomy floorplan in this home, which features the functional and aesthetic warmth of a stone fireplace.

Washington

"I truly enjoy my solid cedar Justus home and highly recommend this type of cedar home to anyone who aspires to live within the walls of quality and casual elegance."

R.&L. B.,WA

Washington

The loft in the Zodiac is an ideal place for a study.

Washington

Making a grand entrance: the owners of this customized Chalet Star added a sunroom off the entry facade.

CONTEMPO PROW STAR

*I*f you like the boldness of the Prow Stars but prefer a lower roof profile – or if building regulations dictate it – consider our Contempo Prow Stars. These spacious homes have wings on one or both sides of the prow and range in size from 1,515 to 3,474 square feet. Any Lindal home can be built on a slab, crawl space, permanent wood foundation or a basement.

Alaska homeowners designed this customized Contempo Prow Star to sit above a walkout daylight basement.

Easygoing style: enjoy a cup of coffee while taking in the view from the glass-front living room.

GALAXY

■ 3 Bedrooms
■ 2 1/2 Bathrooms
■ 3,474 sq. ft.
■ Overall Size: 74' x 41'

This and other Contempo Prow Star plans are shown in larger scale on pages 193-195.

* The term "Star" indicates that the homes in this series have wings.

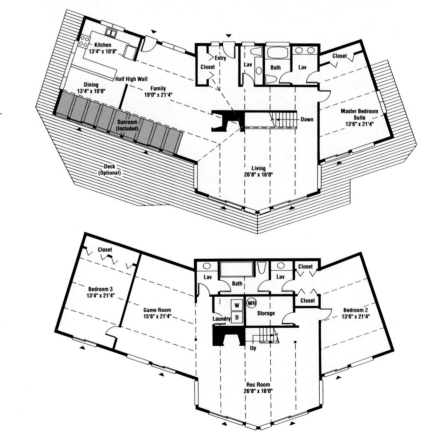

Sliding glass doors from the nine-bay sunroom, living room and master bedroom provide easy access to this large wraparound deck. ➤

"We are most pleased with our Lindal home. It is all we hoped it would be! Not only is it energy efficient, but we didn't have to sacrifice any glass windows to achieve that end. The quality of the Lindal components is A-one, and made the construction of our home smoother than we anticipated."

John and Martha Worley, Lakeville, MA

The knockout kitchen and dining area has an L-shaped eating bar and a second work station with its own sink and cook top.

Skylights and astral windows fill this double-door foyer with an ever-changing play of light.

An aerial view of a spectacular home and boathouse on a lake southeast of Boston.

To brighten up the first-floor family room below the prow, the owners projected it all the way out under the upper deck. ➤

TRADEWINDS

ike the Contempo Prow Stars, these homes have a lower-pitched prow front. But here the wings come together at an angle; and over the ridge where they join, a topknot accommodates an open loft. Think of the topknot as your own private lookout – with an unbeatable view. This home is 2,283 square feet and the other is 2,042 square feet.

The homeowners situated their Windsong so its prow and topknot take full advantage of their spectacular view of Molokai.

WINDSONG

- 3 Bedrooms
- 2 Bathrooms
- 2,283 sq. ft.
- Overall Size: 64' x 54'

This and another Tradewinds plan are shown in larger scale on pages 196-197.

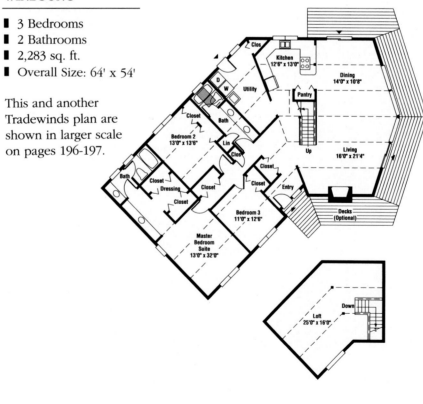

Shades of blue and grey provide a cool, tranquil background to the rich wood furniture and finely-planed cedar timbers.

In the tropics, solid cedar timbers are treasured for more than their beauty alone; they resist termites, rot and decay as well.

Casement windows welcome a gentle breeze into this polished kitchen, with its fieldstone floors and complementary counter tops. ➤

CHAPTER FIVE

TRADITIONAL HOMES

*T*here aren't many opportunities in life to enjoy the finest of yesterday and today. But a Lindal traditional is one of them. These are homes with a heritage. And we make them with the quality materials and craftsmanship befitting their rich architectural history. Yet Lindal traditionals also have a sure sense of the present; their floor plans and amenities reflect the latest and greatest thinking in home design. Create a landmark in your own time with a Lindal traditional, and you'll have a home worth handing down for generations to come.

HERITAGE

The hallmark of our Heritage series is its distinctive gambrel roof, which wraps down to form most of the upper floor's outside walls. The high "R" value of their massive gambrel roofs makes Heritage homes especially energy efficient. Both plans have two full floors – a cost-effective way to build in lots of living space. One-story wings add even more space and graceful lines to their design.

New York

The dining room is dressed up with beloved family heirlooms – silver, crystal, porcelain and fine furniture.

Illinois

The bed is nestled snugly into the gambrel roof line of this second-floor master bedroom.

YORKTOWN

- 3 Bedrooms
- 2 3/4 Bathrooms
- 2,148 sq. ft.
- Overall Size: 47' x 34'

This and another Heritage plan are shown in larger scale on pages 198-199.

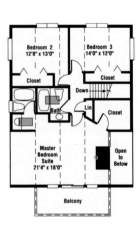

"We love this home. Every part of it! We recommend your company to anyone who wants to live elegantly and expects practicality."

B. & B. deV., NY

Fall colors enhance this home nestled in the green gardens of the historic Hudson Valley. ➤

COLONIAL

Our Colonial series also features a gambrel roof. But it differs from the Heritage by cropping the second floor to a loft. This opens up the downstairs living area and makes way for a cathedral ceiling, two-story windows and open beams. The Mayflower pictured on page 91 features a gambrel plus a one-story wing; whereas the Pilgrim shown on this page has two gambrels intersecting at right angles.

New York

This family bought the 1,000-square-foot core of their Colonial in 1978 with every intention to expand. And they did it with craftsmanship and care, adding the second intersecting gambrel unit on the left and, later, the one-story wing on the right.

New York

Before: the original 1,000-square-foot home in 1978.

PILGRIM

- 3 bedrooms
- 2 1/2 baths
- 2,313 sq. ft.
- Overall Size: 37' x 48'

This and another Colonial plan are shown in larger scale on pages 200-201.

California

Since their site dictated a second-floor entry, these innovative homeowners situated their main living areas on the same level.

"We would recommend Lindal Cedar Homes to anyone who wishes to have a premium quality home... Lindal is the finest of companies with excellent quality materials which have, in every respect, exceeded our expectations."

M. and F. L., Jamestown, NY

The owners built this Mayflower on a full basement and faced it with local stone to complement their grand fireplace. ➢

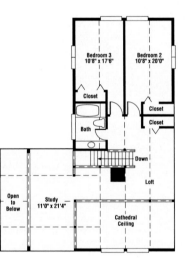

LIBERTY

The focal point of our Liberty series is the two-story picture window dormer, which opens up the center of the massive gambrel roof and floods the interior with light. One plan is 1,608 square feet and the other is 2,075 square feet. Cathedral ceilings rise up to 18 feet over the central living or family room. Upstairs, an open gallery connects the two bedroom wings.

In this highly customized Liberty home in the Rockies, heavy cedar shakes and local stone complement Western red cedar.

CHESAPEAKE

- 3 Bedrooms
- 3 Bathrooms
- 2,075 sq. ft.
- Overall Size: 45' x 27'

This and another Liberty plan are shown in larger scale on pages 202-203.

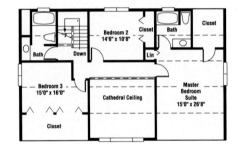

"We have to be the most satisfied customers Lindal can possibly have, and your dealer is a friend we will always have."

Gene and Denise Fiorot, Croton-on-Hudson, NY

The cathedral ceiling maximizes the open ambience of the living room; above, the open gallery connects the upstairs bedrooms.

The two-story dormer, with its window wall, adds interest to the gambrel roof line and illuminates the living room. ➤

FARMHOUSE

There's a rustic appeal to our traditional Farmhouse design. A steeply pitched roof is dormered across the front to expand useful living space upstairs. The most popular feature, however, is probably the covered veranda, which can be designed to your individual configuration. Farmhouse plans range in size from 1,719 to 2,368 square feet.

For the owners of this Farmhouse, it's a pleasant toss-up whether to eat indoors in the cozy dining room or outside on the covered veranda.

Every Farmhouse should have a fireplace; local stone makes this one memorable.

PIONEER

- 3 Bedrooms
- 2 1/2 Bathrooms
- 1,860 sq. ft.
- Overall Size: 40' x 27'

This and other Farmhouse plans are shown in larger scale on pages 204-206.

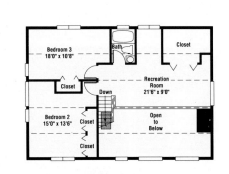

"I like the country atmosphere that my log home has. Also, I dearly love my covered porch on three sides of the house. And the dealer has been very helpful – a good friend."

Vada Wheelock, Traverse City, MI

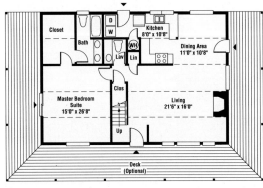

This home was built in round log specifications with optional round log look corners. ➤

NEW ENGLAND

The gambrel roof is the dominant architectural element of this traditional home. Its front facade features a central entry with symmetrical bow windows on both sides and eyebrow dormers above. New England plans range in size from 2,029 to 3,054 sq. ft. All three plans include a full second floor— an economical way to create more living space.

Palladian windows and hardwood floors provide an elegant setting for the formal furnishings in this New England house.

A two-story sunroom adds a special living area that is at home with the traditional architecture.

WHALER

- 4 Bedrooms
- 2 1/2 Bathrooms
- 2,293 sq. ft.
- Overall Size: 37' x 32'

This and other New England plans are shown in larger scale on pages 207-209.

A bird's eye view shows how careful siting pays off. See siting tips on pages 112-115.

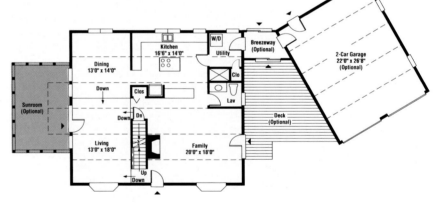

Traditional clapboard siding was the Lindal specification of choice for these Whaler homeowners. ➤

CHAPTER SIX

SMALL TREASURES

Our small homes are living proof that scaling down doesn't have to mean settling for less. A home of minimal square footage can have maximum impact when you  *know how to make the most of every square foot. And we do. The big differences in our small homes: post and beam construction allows a refreshing sense of openness, and smart floor plans embrace a lot of living in a relatively modest space. Whether you live in your Lindal year-round or use it as a leisurely, low-maintenance retreat, we think you'll agree: good things really can come in small packages.*

PROW

🐿

Prow homes have the same captivating features as their larger counterparts in the Prow Star series (pages 178-184): a soaring cathedral ceiling and a panoramic view through a prow of glass. This series simply offers smaller, more economical versions, ranging in size from a modest 766 square feet to a generous 2,300 square feet. If it makes sense to think small now, it's good to know you can easily add a wing later.

"My wife and I both love the design of our house. The huge cathedral ceiling in our living room/dining room makes us feel that we're in a house twice the size."
Paul Carr, Kennebunkport, ME

This solid cedar Prow at Mt. Tahoe is large enough to welcome a house full of skiers.

VICTORY

- 3 Bedrooms
- 2 1/2 Bathrooms
- 2,300 sq. ft.
- Overall Size: 32' x 48'

This and other Prow plans are shown in larger scale on pages 210-212.

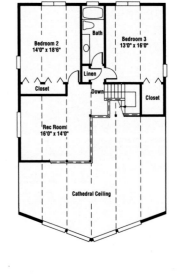

Whatever your view, a Prow can make the most of it 365 days a year.

A brick chimney and stone foundation fit right in with the Western red cedar exterior of this home.

CHALET

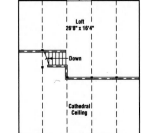

*O*ur Chalet series features the same highlights that make the larger Chalet Star series (pages 187-192) so popular: a cathedral ceiling reaches for the sky and a window wall brings in the great outdoors. Yet these smaller Chalets, which range in size from 1,173 to 1,600 square feet, offer an attractive alternative in size and price. To add variety, and space, one plan has a full second floor.

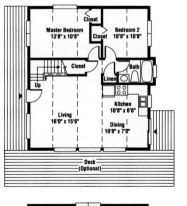

GRENOBLE

▌ 2 Bedrooms
▌ 1 Bathroom
▌ 1,187 sq. ft.
▌ Overall Size: 27' x 30'

This and other Chalet plans are shown in larger scale on pages 213-214.

Stairs lead to an open loft, which is ideal for a study or extra guests.

The enterprising owners of this Chalet raised the walls of its second floor and brought forward the three central modules of the front.

SUMMIT

The popular appeal of our Summit series begins with its striking looks. Its distinctive roof rises from gently sloping sides to a soaring chalet ceiling at the roof's pinnacle. Inside, the sense of drama is heightened by a panoramic prow front. One Summit has an open loft upstairs, while the three larger plans have full master bedroom suites with dormers.

Castle Rock Mountain, near Purgatory Ski resort, provides a magnificent backdrop for this vacation home of a Florida family.

TETON

- 3 bedrooms
- 2 baths
- 1,476 sq. ft.
- Overall Size: 32' x 36'

This and other Summit plans are shown in larger scale on pages 215-216.

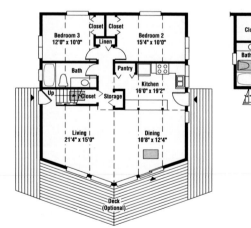

"After two years we are most happy with the house, especially in the winter, when we feel so snug with the fireplace going and the cold and rain outside."
Bill Noland, Allyn, WA

The distinctive roofline is the signature of our Summit series. ➤

CONTEMPO PROW

This series offers the same main attractions of the larger Star series (pages 193-195) – namely, a prow front of glass and a vaulted ceiling with open beams. But here you'll find smaller, more economical floorplans, ranging in size from a compact 1,173 square feet to a more spacious 1,887 square feet. Some plans come with a topknot. And, of course, you can add a wing whenever you wish.

The all-glass front of this Contempo Prow gives its Maui owners a panoramic view of Lanai and Kahoolawe.

Hawaii

Alaska

A frozen lake not far from Anchorage is the setting for this solid cedar Contempo Prow; the topknot takes full advantage of the view.

Hawaii

Rattan furniture is well-suited to the casual decor of this Hawaiian home.

NEPTUNE

- 3 Bedrooms
- 2 Bathrooms
- 1,887 sq. ft.
- Overall Size: 43' x 39'

This and other Contempo Prow plans are shown in larger scale on pages 217-218.

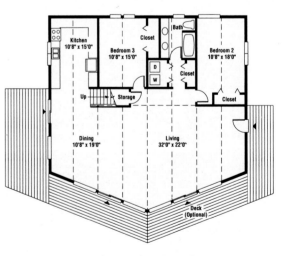

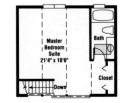

VIEW

*T*he name of this series almost speaks for itself. Simple but striking, our View homes bring together the best of Lindal in a small package: the beauty of cedar, the drama of open beamed ceilings, and the invitation to the outdoors offered by the window wall. View homes can be built on a slab or a full or daylight basement. In fact, because so many of them are built on hillside sites, one plan is specifically drawn with a daylight basement.

"We are glad we chose Lindal. After a hectic, busy day, it is so nice to be welcomed by a relaxed, cordial atmosphere. I would be pleased to recommend Lindal homes to anyone."
Sharon Wood,
Half Moon Bay, BC

WAIKIKI

- 2 Bedrooms
- 1 1/2 Bathrooms
- 1,729 sq. ft.
- Overall Size: 32' x 37'

This and other View plans are shown in larger scale on pages 219-220.

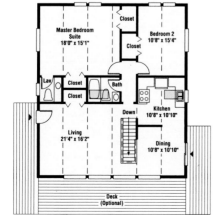
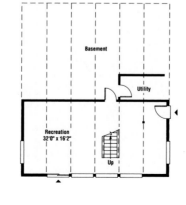

Astral windows add even more light and spaciousness to this view home in Alberta, Canada.

PANORAMA

🐿️

*T*ake advantage of a wide view site with a Panorama plan. The entire front wall of glass is designed to take in the scenic grandeur all around you. The roomy, U-shaped Monaco plan, pictured here, has an open loft upstairs. A smaller version, the Acapulco, has a topknot that raises the height of the open loft and creates greater space for a master bedroom suite.

"Three thousand miles from your factory to our site, down a narrow three-mile dirt road, delivered on time, within the price quoted. Thanks to you, (Chuck Reid, the dealer) and the builder."
Shirley and Christopher Sherrill, Wakefield, NH

The blue-and-white color scheme makes this bedroom as fresh and inviting as a summer day.

MONACO

- 3 Bedrooms
- 1 3/4 Bathrooms
- 1,555 sq. ft.
- Overall Size: 43' x 37'

This and another Panorama plan are shown in larger scale on page 221.

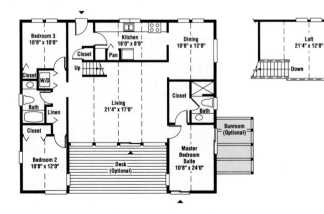

A wall of glass gives this Vancouver Island home a striking indoors-outdoors beauty. ➤

GAMBREL

🐿

*I*n essence, our Gambrel series features the traditional style of our Heritage and Colonial homes – without the wings. All have gambrel roofs, which wrap down to form much of the walls of the second floor. Two plans are Heritage in style, with two full floors. The third plan – Colonial in style – has a cathedral ceiling over the living and dining area.

Patio doors lead to a second-floor balcony off this custom Gambrel, which has its main living quarters on the second floor, too.

"Our coastal hilltop location leaves us exposed to the worst weather from the Atlantic. But the energy efficiency of the design has provided us with a snug, warm home with minimal energy cost."
Carl Canzanelli,
Newtonville, MA

LEXINGTON

- 3 Bedrooms
- 1 1/2 Bathrooms
- 1,614 sq. ft.
- Overall Size: 27' x 31'

This and other Gambrel plans are shown in larger scale on page 222.

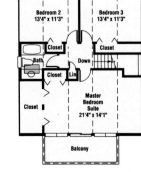

Texas

Texas

"When 60-mile-per-hour winds sweep across the flats, it's a cozy feeling to be in the house. It's airtight to drafts and insulated, so you are warm in winter and cool in summer without it costing an arm and a leg in electricty."
Nancy Crowe,
Fort Davis, TX

POLE

🐿️

ole homes are the ideal solution to challenging hillside sites, beaches where flooding may occur—anywhere you want to raise your home above the ground. Aesthetically, these homes have a special character, a poetic stance that sets them apart. A sheltering roof with wide overhangs and a wraparound deck make Pole homes perfectly suited to sites from the Gulf States to tropical zones worldwide.

QUADRA

- 3+ Bedrooms
- 2 Bathrooms
- 2,304 sq. ft.
- Overall Size: 48' x 48'

This and another Pole plan are shown in larger scale on page 223.

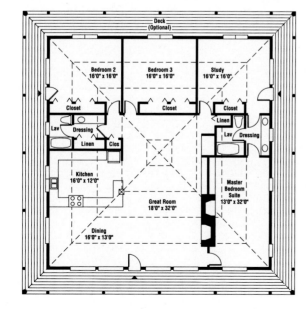

A photograph of the exterior of this home is shown on pages 98-99.

Hawaii

Pole homes are ideally suited to the tropics; this one is at home on the north shore of Oahu.

CHAPTER SEVEN

HOME PLANNING IDEAS

*N*ow that you've seen some living examples of Lindal homes, we'd like to turn the spotlight on you. This section is devoted to helping you ask, and answer, all the right 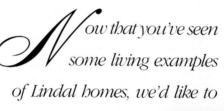 questions as you plan your custom Lindal. We'll share our expertise with you, from basic siting and design principles to planning kitchens, baths, windows and much more. For most people, making these choices is one of the most exciting steps in building their own home—exceeded only, of course, by the finished results.

BEFORE YOU START

Our Lindal homebuyers tell us there's a special joy to living in a home that you designed yourself. And it's surprisingly easy when you get off to the right start. That means taking time to think about your *wants and needs*, your *site* and your *budget* – the three things that help you shape your dream into reality.

Start With A Wish List and Notebook

Now's the time to gather the family around and start a list. One bedroom for each family member? An "adults-only" wing? A kitchen you can entertain in? A sun-filled breakfast nook? A computer center for the entire family? A studio over the garage for your artist-in-residence?

Next, take a practical look at your list to distinguish between "wants" and "needs." You may want one bedroom per family member plus a guest suite for the grandparents, but is it feasible? Only you can say. Making these distinctions will help you prioritize your needs and stay within your budget.

Now you can get more detailed. We recommend that you keep a design file or notebook for each area of your home. You may already have a good start from browsing through magazines and books. Include articles and photos, product literature, color swatches and paint chips – anything and everything that helps bring your vision to life and communicate it to others.

Think About Your Architectural Look

Whether it's a traditional for your historic neighborhood – or a prow-front contemporary for your view site, now's the time to think about your home's architectural personality. We're ready when you are; this book is filled with dozens of variations on our four major families of architecture: Classic (pages 153-177); Contemporary (pages 178-197); Traditional (pages 198-209); and Small Treasures (pages 210-223).

Consider the number of levels you want, according to your site and lifestyle. At the same time, you'll want to choose the roof style, since it tops off your architectural look. Remember, you can use any roof style – including gable, hip, gambrel, saltbox, Cape Cod, shed, and more – on any Lindal home.

An architect typically spends many hours working out this stylistic "fit" between client and building site, but you can do it yourself – perhaps more completely. Of course, you may want to start with an architect and bring your preliminary plans to Lindal for completion; some of our homeowners do. However you want to work, we're happy to be part of your team.

Get to Know Your Building Site

The best homes are a synergy of building and site. So get acquainted with your site. (We recommend a camping trip!) Visit it at different

Expand your wish list into a design file or notebook. Get all your wants and needs down on paper, adding ideas from books and magazines.

Overall Requirements:
- The traditional New England look is for us: Gambrel roof, bow windows, shutters.
- Two stories.
- But we want lots of natural light. How about a sunroom? Skylights?
- Must have a deck off the kitchen for the barbeque and outdoor furniture.
- Cedar on the outside.
- A combination of cedar and drywall inside
- Kitchen is important! U-shape is best. must be large enough for all of us to eat together.
- Want lots of storage: pantry, walk-in closets, linen, a place for folding laundry; ironing—and more storage.
- Budget = $150,000 plus lot (we have it!!!)

LIST OF ROOMS

Room	Used by	Size	Sunlight	Next to
Cook	All	12×11	Am	Eating
Eat (kitchen)	All		Am	Kitchen
Eat (dining)	All	20×12	Pm	Living
Eat (out)	All		Pm	Living
Sleep	D&M	16×21	Pm	Bath
Sleep	T	16×11	Am	Bath
Sleep	J	13×11	Am	Bath
Bath	D&M	6×10	Am	Bedroom
Bath	T&J	5×7	Am	Bedroom
½ Bath	All	5×5		Kit/Eat
Living/quiet	All	16×21	Pm	Dining
Living/noise	All	16×21	Am	Kitchen
Den/quiet	D	16×14	Pm	

times of day – and night. Discover the effect of changing seasons. Experience the ebb and flow of light, the path of the sun, and the changing scene around you; it will help you create a home that makes the most of its surroundings.

While walking your property, try and visualize what style of home will best suit your site as well as your wish list. Choose the most likely spot for your home, sit down – and give yourself plenty of time to mull over the possibilities.

Develop a Site Plan

You may already have a survey of your site. If not, you can draw your own working site plan, sketching in the property lines, ground slope, roads, water and gas mains, nearby buildings and other important features, from trees to bodies of water. Use arrows to indicate "best views."

At this point, you'll probably need to investigate:
- Zoning
- Covenants
- Setbacks
- Height restrictions
- Building codes for snow and wind loads
- Accessibility for trucks and equipment
- Soils and drainage/perc tests
- Sewers or septic systems
- Electricity, gas, water, phone and cable TV

These factors will influence the design of your home and its site location. They'll also have an impact on cost.

Consider the Sun

Be sure to think of the sun in relation to your site. Do you like it to wake you up in the morning? Or do you prefer to greet it once you're awake – in the sunroom over coffee? Do you look forward to watching the sunset from your living room? Such preferences should influence your floorplan. And with high fuel costs sure to get higher, it makes economic sense to capitalize on the sun's free energy. With passive solar in mind, be sure to indicate "North" on your plan.

Write Your Own Program

After studying your needs and wants, site and budget, an architect would develop a written program. You can, too. It consists of three parts:

1) Overall Requirements: Describe the look and feel of your ideal home.
 - Architectural style: Traditional or contemporary?
 - Number of stories: Sprawling ranch or multilevel?
 - Roof type: Gable, gambrel, hip, or other?
 - Interior style: Open and informal or contained and formal – or a combination?
 - Lifestyle: Big family, young adults, retirees?
2) Individual Requirements: List the wants and needs of each family member.
3) List of Rooms:

Get detailed here, listing each proposed room, user(s), size and special needs.

Congratulations!

You've completed the most important part of the planning process – turning a dream into a specific list of requirements. Now you can move ahead with confidence. You'll be able to evaluate the plans in our design library (pages 153-223) according to your personal criteria. Perhaps one will suit you to a "T," or require only minor modifications. If not, don't hesitate to let your imagination soar and design a one-of-a-kind based on what you've learned about you and your dream home. Any questions? Your Lindal dealer is here to help.

Individual Requirements

DAD
- Quiet place to work at home, preferably with view

MOM
- Big kitchen
- Large dining room for big groups
- Walk in closets

Trina & John
- Separate rooms
- Lots of storage
- Two sinks
- Makeup vanity

North

June 21 (Long Days)

Sept 21 / Mar 21 (Equinox)

Dec 21 (Short Days)

Think about the sun's path during the day; how does it relate to your site – and you?

In the summer, an overhang can block the high summer sun; in the winter, when the sun is lower, its light streams in – and is a welcome visitor.

Summer   Winter

FROM SITE TO FLOORPLAN

With your site plan and written program in hand, you're ready to explore the relationship between actual spaces in your home. We recommend the same technique architects use: play with circles that represent these spaces.

In each circle, jot down a letter that identifies the room for you ("B" for bedroom, "K" for kitchen, and so on). Now play with your circles. Make loose sketches, and don't worry about scale. The spatial relationships are all that matter here. This exercise will help you find the best locations for each room. For example, if you want morning sun in the kitchen, you'd place the kitchen to the east.

Next: Loose Line Drawings

Before you actually start line drawings, browse through the rest of this chapter for more design guidelines. When you're ready to move from generalities to specifics, here's how:
1. Circles: Sketch the relationship between spaces.
2. Initial drawing: Spaces become rooms.
3. Preliminary line drawings: Rooms become a floorplan.

Drawing to Scale Is Easy

Congratulations again! You're ready to start drawing your floorplan to scale on the grid paper included with this book. Page 152 has more instructions.

An important note: You may find the design process so much fun that you want to do it all yourself. And with these guidelines, you can. But remember that, at any point, you can turn the project over to your nearest Lindal dealer, who is trained in design and who can give you a complimentary feasibility and cost analysis.

Getting Down to Floorplans

Traffic Patterns

Your main entry should allow you to enter any room in the house without passing through other rooms. Ideally, it gives guests quick, easy access to the living room. And if it's close to the kitchen, you'll save steps when you greet them.

The secondary entry should lead directly to the kitchen. If there's a place between this entry and the kitchen for coats and paraphernalia, so much the better.

Think about the daily routine of each family member and "walk it through" the plan you're considering to see how it works. Give special attention to such daily chores as answering the door and carrying in groceries.

Zoning for Activities

It's important that your floorplan honors your family's three basic activities – working, sleeping and entertainment – by situating them in harmony. Ideally, the noisiest zones, usually the work areas, are grouped together and well away from sleeping areas. If you have young children and entertain often, these respective areas should be separated, too.

In residential neighborhoods, it's often best to locate the living zone toward the rear of the house, away from street noises and passersby. Your site's elevation, view, surrounding landscape and other factors will

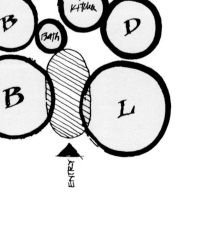

1 Using circles, sketch the relationship between spaces.

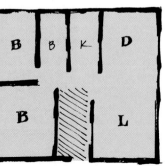

2 Now, turn those spaces into actual rooms.

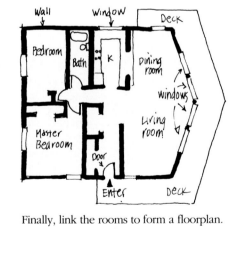

3 Finally, link the rooms to form a floorplan.

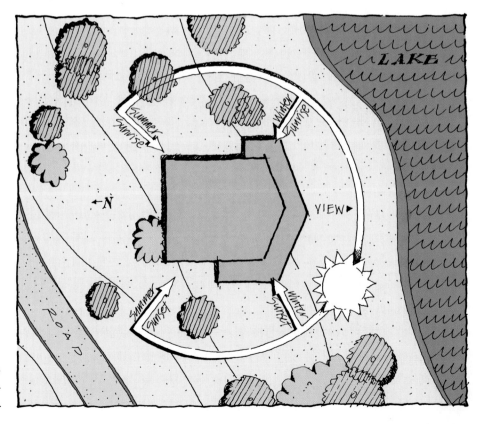

This site plan shows the home on the site and its relationship to sun and views.

also enter into this decision.

Each bedroom should have access to a bathroom within the sleep zone and, where possible, be buffered from other bedrooms for the sake of privacy and quiet.

Closets, staircases and bookshelves all can serve as sound barriers to keep noise from filtering into bedrooms.

These same zoning considerations apply to a two-story floor plan. In addition, you'll want at least a half-bath on the main level.

Main Entry

Your main entry greets guests and gives the first impression of your home. As such, it should be inviting and well-defined, leading directly into the formal living area. Yet it should also be private enough that the entire house is not on view from the open door. A large entry closet is a real convenience.

Kitchen and Secondary Entry

Life is easier when you have access from the garage to the kitchen; it simplifies such regular chores as unloading groceries and taking out the trash. If the entry is through a utility or mud room and has two doors for an airlock, you'll save on heating and air conditioning.

The kitchen should be convenient to the dining room, family room and outdoor patios and decks. If the cook likes to stay in touch with family members and guests, or if you have small children, it's a good idea to have the kitchen open to the family room. A pass-through provides easy access and a degree of separation. See the next four pages for more information on planning your kitchen – including actual kitchen layouts.

Bedrooms and Baths

Bedrooms should have two walls uninterrupted by doors or low windows, giving you the flexibility you need to arrange furniture. Locate closets between rooms to cut down on noise transmission.

Children's bedrooms should have ample space for playing and a desk for studies and hobbies.

The door to the master bedroom should be placed so you needn't enter the sleeping area to reach the bath or closets. One smart solution includes an area opening off the master bedroom and encompassing the closet, bathroom and dressing room in three separate, but connected spaces.

For more information on bathroom design, see pages 120 and 121.

A main entry that allows a partial view of other areas builds expectations but maintains privacy.

Your main entry should let you enter most rooms and go upstairs without passing through other rooms.

A bath en suite turns the master bedroom into a private retreat.

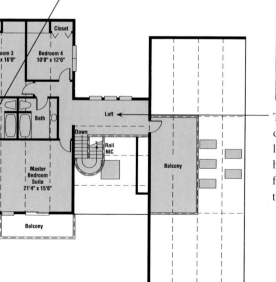

Opening the dining room to your living and family areas adds visual space to all of them.

The stairwell, loft and balcony look onto views of the living areas below - and both levels benefit spatially from the visual openness they share.

Zoning

- Work Areas
- Entertainment Areas
- Sleeping Areas

Two-story floorplans are especially good for zoning. Bedrooms are usually located on the second floor. If so, be sure to include at least a half-bath on the main level.

ONE OF A KIND #9

- 4 Bedrooms
- 3 Bathrooms
- 3,344 sq. ft.
- Overall Size: 59' x 56'
- Master bedroom on second floor

KITCHEN SUCCESS: PLANNING IS THE KEY INGREDIENT

*T*he kitchen has always been the heart of the home. And it's as true as ever today. Many of our fondest memories revolve around the foods, fragrances and friendships cooked up in this room. At Lindal, we think it's a special place that deserves special attention.

The best kitchens start with homeowners who have done their research and know what they want. With the cost of kitchens today, careful planning is essential.

Do the kids do their homework in the kitchen? Do guests always seem to congregate there, too? What else goes on in your kitchen besides cooking? These are just a few of the questions you'll want to ask yourself.

The trend today is toward the kitchen as family living center – an area spacious enough for a planning desk as well as food preparation and, of course, family dining. Many such kitchens are wide open to dining and living areas.

Let your kitchen express your personality and style. But before you start, take advantage of some tried-and-true guidelines.

The Classic Work Triangle

This practical bit of spatial planning locates the sink, cooktop and refrigerator to form a triangle within the kitchen. The goal? To reduce the steps between these work centers – and keep extraneous traffic outside. No matter how large your kitchen is, the total distance of this triangle should be no less than 12 feet, no more than 22 feet. The distance between appliances should be no less than 4 feet, no more than 7 feet.

Each of the three areas should make up its own work center, with counters on both sides of the sink and cooktop, and a counter on at least one side of the refrigerator. Put the sink in the middle.

However, there's often the need for more than one cook in the kitchen. Then it may make sense to add more work centers, and break the triangle rule.

Massachusetts

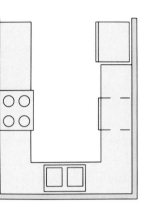

A free-form eating area adjoins the island in this crisp-looking kitchen.

New Jersey

U-Shaped
With three walls, the U is the ideal kitchen layout. Each appliance in the triangle has its own counter, and the sink is usually at the base of the U. This dead end eliminates traffic in the work triangle. Two cooks can use the U comfortably.

Single Wall
This layout is most common in vacation and compact homes. Although it does not allow a work triangle, the distances between work centers are typically short. Single-wall layouts work best with the sink in the center.

L-Shaped
Two walls at right angles form an L, which accommodates an efficient triangle that lets traffic pass without interrupting the cook. The L consumes less space than the U and is almost as efficient.

Corridor
The corridor kitchen has two parallel counters, 4 to 5 feet apart. It adapts well to an open kitchen, with a row of base cabinets serving as a room divider. One drawback: traffic can interrupt the work triangle.

Peninsulas
Unlike a free-standing island, a kitchen peninsula is attached at a right angle to one wall of the basic layout. Here, it makes a U out of an L.

Islands
Kitchen islands work well in large U and L layouts; they help minimize the distance of the work triangle in big kitchens. They often double as eating bars, too.

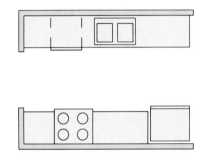

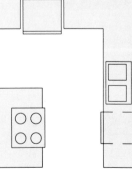

An authentic oldtime gas range, pine furniture and colorful accessories bring this country kitchen to life. ➤

JOANN HOFFARD'S KITCHEN

Joann Hoffard is an interior designer who lives in the frontier town of Auburn, California in the rolling foothills of the Sierras. She and her husband, Bud, chose a Justus Prow Star, which expresses the love they share for beautiful wood. The entire home is a tribute to Joann's artistic decorating, and nowhere is her engaging mix of creativity and practicality more evident than in her kitchen – as you can see by the photos and comments she has generously shared.

"We love the rich, warm natural wood textures and colors, and never get tired of the compliments from visitors. The modifications we made to the basic plan were minor, but very important to us. We feel that we have a tailor-made home, and are very proud of it."

Joann Hoffard, Auburn, CA

A view across Joann's kitchen, with its island to the dining room.

Details that make a difference:

Roll-out shelves for linen storage.

Built-in gift wrapping center.

Magazine rack keeps light reading handy.

Behind herbs and spices is more storage for larger containers.

Kitchen Tips

- Be sure the sink won't be blocked when the dishwasher door is open.
- The sink should have the most counter and storage space on both sides.
- The refrigerator door should hinge away from the sink and cooktop.
- Locate food storage near the refrigerator.
- Store pots, pans, utensils and spices near the stove.
- You'll need at least 8 to 10 feet of counter space, excluding appliances.
- Space between counters should be at least 36 to 48 inches – 60 inches if cooking is often a two-person operation.
- Consider a second sink for salad preparation or a wet-bar outside the triangle.
- Since the oven is used less than the cooktop, it can be located outside the triangle.
- A microwave can be handy outside the triangle, too.
- Lighting is a highly specialized field; consult the experts.
- Do your homework on appliances and equipment. Learn as much as you can about brands and features – then buy the best you can afford.
- Functional storage is essential to a convenient, efficient kitchen. Plan for lots of it – then double your estimate. You'll find you never have too much.

Red tile and chrome appliances jazz up the black-and-white color scheme in this open kitchen and dining room design.

A free-hanging light fixture, made of cedar, illuminates the island in this already bright and spacious kitchen.

Skylights over the roomy island fill this kitchen with light.
The sunroom lies just beyond the open pass-through.

THE BATHROOM: FROM BASIC TO LUXURIOUS

*Y*our home planning files are probably filled with ideas for the perfect bathroom: a sybaritic spa, his-and-hers dressing areas with custom-built-ins, a window wall with a view. You can design any bathing beauty you want in a Lindal home. Of course, budgets come into play. Our guidelines will help you get the most bathroom and beauty for your money.

Bathroom Tips

- Whenever possible, locate bathrooms on outside walls to take advantage of natural light and ventilation. Or use a skylight – preferably a ventilating one.
- To maximize space, consider lavish use of windows. How about a garden window? A skywall over the tub? Or a full sunroom? Here's a room where Lindal sun products really shine.
- Mirror small spaces to expand them visually.
- Pocket doors are another space saver.
- Like kitchens, bathrooms benefit from professional lighting expertise.

- Plan ample storage for grooming and bath supplies as well as linens.
- Make one bath work as hard as two by adding privacy walls–or doors– between fixtures. Two people can comfortably share a compartmentalized bath at the same time – a real blessing on busy mornings.
- Add amenities to create a unique bathing retreat: a second sink, separate shower and tub enclosures, a sit-down makeup vanity– and lots of storage. A linen closet is a real plus.

Imported marble in relaxing pastels lines the jacuzzi in this spacious master bath. High astral windows let in natural light without sacrificing privacy.

Rose-colored ceramic tile lines the jacuzzi in this corner bath, which is rich with windows to enjoy the view.

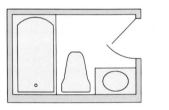

Full bathroom.

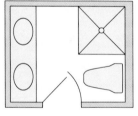

Three-quarter bath.

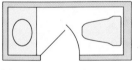

Two-piece powder room.

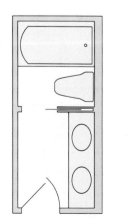

Full bath, compartmentalized, with two sinks and one entry.

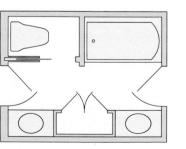

Full bath, compartmentalized, with two sinks and two entries.

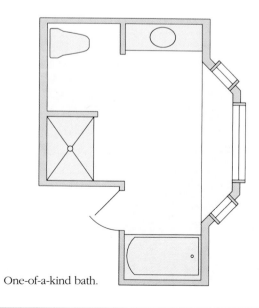

One-of-a-kind bath.

Texas

New York

A black jacuzzi tub and gold accents add to the opulence of this master bath, which is lined entirely with imported marble.

Vermont

Blue tiles and white fixtures give a fresh look to this double-sink bathroom. The narrow cedar door opens to a linen closet.

Gold faucets glamorize the jacuzzi tub in the burgundy-toned bath. Casement windows open for ventilation.

WINDOWS: DESIGNING FOR LIGHT

*M*ost of us gravitate toward open rooms, natural light and a wonderful view. It's *glass* that makes all of this possible, which is why windows play such an important role in the architecture and enjoyment of your home. They not only let in light and air, but also the view – linking the outdoors to the indoors, and making a major contribution to your home's overall appeal and livability.

Vermont

A fan window centered over French doors and twin side lights – all of them gridded – is an elegant way to infuse this living room with light.

Window Tips

- Think about light. Northern light is poetically diffuse, but north-facing windows are energy-losers in winter. Eastern light is soft and pleasing – perfect for morning breakfast. Southern light is best for passive solar gain, but plan for protection when it comes on strong. Western light brings the sunset with it, but needs to be controlled in the summer.
- The most pleasing room lighting comes from windows on two walls.
- Decide where you want light and views, and where privacy is most important.

- Determine where you need ventilation.
- Can you take advantage of passive solar by placing lots of glass to the south? (For more information, see pages 144-145.)
- Think about the exterior look of windows, as well as their interior use. Try for harmony and consistency– inside and out.
- Consider windows to complement your architecture. For example, double-hung windows with grids for traditional homes and large, undivided expanses of glass for contemporaries.

Note: Lindal windows are available framed in clear Western red cedar and/or Douglas fir. Cedar framed windows have a heavier profile than fir frames. Some windows are custom orders, which require a longer lead time to delivery. For more details, check with your Lindal dealer.

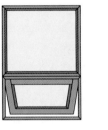

Casement
A tall, vertical window, hinged on one side to open like a door. Excellent ventilation.

Awning (Under Fixed Glass)
A long, horizontal window, hinged to open out from the bottom. Excellent ventilation and good rain protection when open.

Single Hung
A traditional window; the bottom half slides up to open. Optional grids are available. Custom order.

Double Hung
Another traditional. Both upper and lower halves slide independently. Optional grids are available. Custom order.

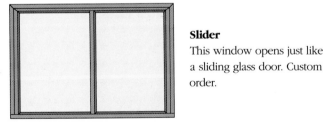

Slider
This window opens just like a sliding glass door. Custom order.

Fixed Glass
Less expensive than opening windows, fixed glass windows provide the greatest energy efficiency and security of all. Available in a wide range of sizes and shapes, from squares and rectangles to triangles, trapezoids, clerestories and astrals, they often are used in conjunction with opening windows.

One-of-a-Kinds
Your imagination is the limit: fans, half-rounds, bows, bays, rounds and octagons. Custom order.

Jalousies
Wood and glass louvers open for ventilation and offer good protection in the rain, but are low in energy efficiency and security. Suitable for tropical climates.

Important: Windows have an enormous impact on the energy efficiency of your home. For more information, see page 142.

A natural accent: Western red cedar frames the fixed glass in the window walls of this living room.

Design your windows to frame your view. Notice how the top triangular windows echo the lines of the roof.

A fan crowns a trio of opening casement windows in a light-filled room.

EYE-CATCHING OPENINGS

As you can see here, windows are just the beginning when it comes to taking in a view and opening your home to light and ventilation. Take advantage of our wide selection of doors, skylights and three-dimensional garden windows, too.

Doors: Sliding Glass, Patio and French

Choose glass doors to extend living out-of-doors – from the kitchen to the barbecue, from living room to deck, from your bedroom to a secluded balcony. Consider interior uses, too: as an entry to the sunroom or indoor swimming pool, or around a central atrium. *Note: For security and energy reasons, French doors are best used inside the home.*

Garden Windows

A garden window can bring the delights of a miniature greenhouse into your kitchen or bathroom. This wood framed, three-dimensional window has a shelf that gives your plants their own place in the sun and lets you grow your favorite herbs or orchids all year round. Vented side panels open to catch summer breezes.

Skylights

Imagine lying in bed and gazing at the star-filled sky. A skylight makes it possible, and it can create magic in other areas in your home, too: over your entry, in a long corridor or small bathroom, over a stairwell – anywhere you want added drama, light and ventilation.

Like windows, skylights can be fixed or opening. The Velux® skylight is fixed but has a ventilation flap; the Velux® roof window is available in two opening models: one pivots completely from the middle, so both surfaces can be cleaned from the inside. The other opens at the bottom for ventilation without removing the screen.

Note: All of our products are tops in energy efficiency and wood framed for the ultimate in richness and warmth. Naturally, they blend beautifully with the cedar exterior of any Lindal home.

A three-dimensional garden window with a shelf and vented side panels.

Two fixed Velux® skylights fill this kitchen with natural light.

A pair of sliding glass doors opens in one easy, gliding motion. The skylight provides more natural light.

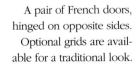

A pair of French doors, hinged on opposite sides. Optional grids are available for a traditional look.

Swinging patio doors hinge together in the center and fold back inside. Optional grids are available. ➤

HARDWOOD FLOORING

*C*lassic hardwood floors are the ideal combination of practicality and beauty for today's demanding lifestyles. Their good looks, warmth and character are unparalleled. And they're hardworking too: durable, long-lasting, easy to maintain.

Hardwood Flooring Tips

- Choose unfinished hardwood so you have an unlimited choice of color and finish.
- Basic finishes include bleached or pickled, smooth or textured, shiny, semi-gloss or satin.
- Final finishes include Swedish finish, polyurethane or waxed finish.
- Hardwoods can cost less than good quality carpeting but, unlike carpeting, they will last the life of the home with proper care.

- With modern finishes, dust mopping and occasional wet mopping are usually the only maintenance required.

Lindal Hardwoods

We mill our own premium Ontario Hard Maple and Northern Red Oak at our hardwood plant in Ontario, Canada. Within each species, we offer two grades, which differ in color. Both grades supply the same high strength, durability and long life.

The pale finish on these maple hardwood floors suits the soft colors and subtle decor of the living room.

Rock-hard maple hardwood, with its amazing durability, is a smart choice for this snazzy black-and-white kitchen.

Ontario Hard Maple: Rustic

Ontario Hard Maple: Prime

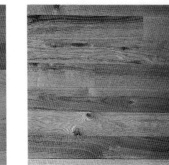
Northern Red Oak: Rustic

Northern Red Oak: Select

"Our Hawaiian lanai, is a dream. We watch the activities such as fishing tournaments (75 to 100 boats!), canoe racing, water skiers, jet skiers, parasailing and the cruise ships' twice-weekly visits."

Mr. and Mrs. Marcus Rosehill, Kailua-Kona, HI

A deep roof overhang shelters the lanai of this home from the sun and torrential downpours.

In areas where insects can be irritating on hot summer evenings, a screened-in porch solves the problem beautifully.

THE GREAT OUTDOORS: DECKS, LANAIS AND SCREENED PORCHES

*T*hese architectural elements let you extend your living space by creating outdoor rooms for relaxing and entertaining *al fresco* style.

You can design your deck to any size and configuration, according to your needs and the topography of your site. Make it multi-level, if you like. A raised deck off a second story becomes a balcony. A choice of deck rails is available. Sometimes a rail with built-in seating is a smart choice: functional and attractive, too.

A screened-in porch is a blessing in hot climates where insects are a problem. These are cool, relaxing retreats that conjure up visions of wicker chairs and iced tea.

Tips for Outdoors

- For beauty and practicality, choose decking, railing and undercarriage in Western red cedar, which resists rot and insects.
- Make sure the planks for your deck are a solid 2 inches thick to protect against warping and twisting - and to prevent the bounce often found in decks.
- Sliding glass, Patio or French doors make it a breeze to serve meals outdoors.

With a view like this of Topanga Canyon, eating out-of-doors on the deck seems the natural choice.

SUNROOMS ET CETERA

One of the most exciting recent developments in residential architecture is the sunroom's coming of age. No longer a greenhouse or an add-on that looks like an afterthought, sun products can be integrated into the design of your home, enhancing its looks – and surrounding you with the beauty of the outdoors 365 days a year.

Does your dream home feature an all-glass entry? A two-story sunroom that reaches to the sky? A lap pool open to sun and stars? Or maybe a sun-drenched room off the kitchen where the family can kick back and relax? Our professional designers can translate any dream under the sun into a reality.

A ten-bay sunroom lets the sun shine in on this 40-foot lap pool in a custom Lindal near Ottawa.

The SunCurve's graceful arches (above) give an entirely different feeling than the clean-lined Straight Eave SunRoom (opposite). Choose the style that suits you.

Note: For plans, specifications and passive solar techniques, see the 28 page Lindal Cedar SunRooms brochure. Constructed of wood and glass, Lindal SunRooms harmonize beautifully with Lindal homes – both in looks and quality of materials and craftsmanship. The wood is premium grade, kiln-dried clear grain Western red cedar – and the glass is thermal and energy efficient.

Choose from Six Sun Products

Straight Eave SunRoom
Overhead glass joins the straight, vertical sides in a wide variety of roof pitches. Cedar mullions are joined in the furniture maker's favorite connection – the bird's mouth joint – for an airtight and watertight fit.

SunCurve SunRoom
Graceful, flowing lines set apart the SunCurve with its satin-smooth laminated arches. The curved glass is uninterrupted by cross bars within line of sight. Each distortion-free glass bay is 3 feet wide.

SkyWalls
Our SkyWalls are designed primarily for kitchens, but they're popular in bathrooms, too. They are designed to sit on a 40-inch base over counters, inset into the roof pitch to a depth of 3 or 6 feet.

SunDormers
Consider a SunDormer for your second story. Imagine glass walls and glass overhead instead of solid wall and roof. Perfect for flooding your bedroom or bathroom with sunshine and sunlight. Available in virtually any size.

SunWalls
Consider a SunWall where you'd like a wide vista of glass, but where overhead glazing is not desirable e.g. hot climates or heavy snow loads. Available in 4 and 6-foot glass heights.

SunCorner
SunCorners are specially designed to open up the home on two sides, with a hip roof to wrap around the corner. The result is a highly customized look, available in nearly any size.

This Straight Eave SunRoom is off the kitchen of the one-of-a-kind home shown on pages 26 and 27. ➢

CHAPTER EIGHT

SPECIFICATIONS

Here's where we get down to the nuts and bolts of Lindal quality, detailing precisely what goes into every  *Lindal custom cedar home – as well as the differences that distinguish each of our four major styles: Lindal, Justus, Clapboard and Round Log homes. Within these specifications, you'll find a wealth of energy-saving options that let you customize your Lindal to your climate, whether you're building for cold New England winters or humid Carolina heat. Solar is one of your options; what could be more resourceful than tapping the free energy of the sun?*

LINDAL MEANS QUALITY

Throughout this book you've seen many of the quality features and attention to detail that make a Lindal home special. Many of the features are standard, others are optional – and still others are strictly homebuyer innovations that we're happy to share.

For your convenience, we've brought together some of the most important features – the details that distinguish Lindal – right here, on these two pages. All are available to you as you begin to choose the details that make a Lindal home *your* home. For more information, consult your local dealer.

Specifications are subject to change. Always consult the current general specification sheet.

Hawaii

"There's simply nothing on the market today that has the quality of Lindal. We love being surrounded by the fragrance and warmth of cedar." Eli Pearly, Kula, Hawaii

Hardwood Flooring.
Maple and Oak. See page 126.

Cedar Framed Windows.
See pages 122, 123, 142 and 143.

Fir Framed Windows.
See pages 122 and 123.

Jalousi
Wood or glass

Garden Windows.
Wood framed.
See pages 124 and 125.

Skylights and Roof Windows.
Wood framed / Velux ®
See pages 124 and 125.

Cedar Framed Sunrooms.
See pages 128 and 129.

Steel Entry
see page 1

Wood Bifold Doors.
Flush, hollow-core, louvered wood or raised panel cedar

Cedar Garage Doors.
8', 9', and 16' sizes.

Casing.
2¼" cedar or Oak casing

Horizontal Laminated Beam
see page

Cedar 2×2 Rail.

Cedar Simple Rail.

Cedar Seat Rail.

Wood Ro Overha
see page

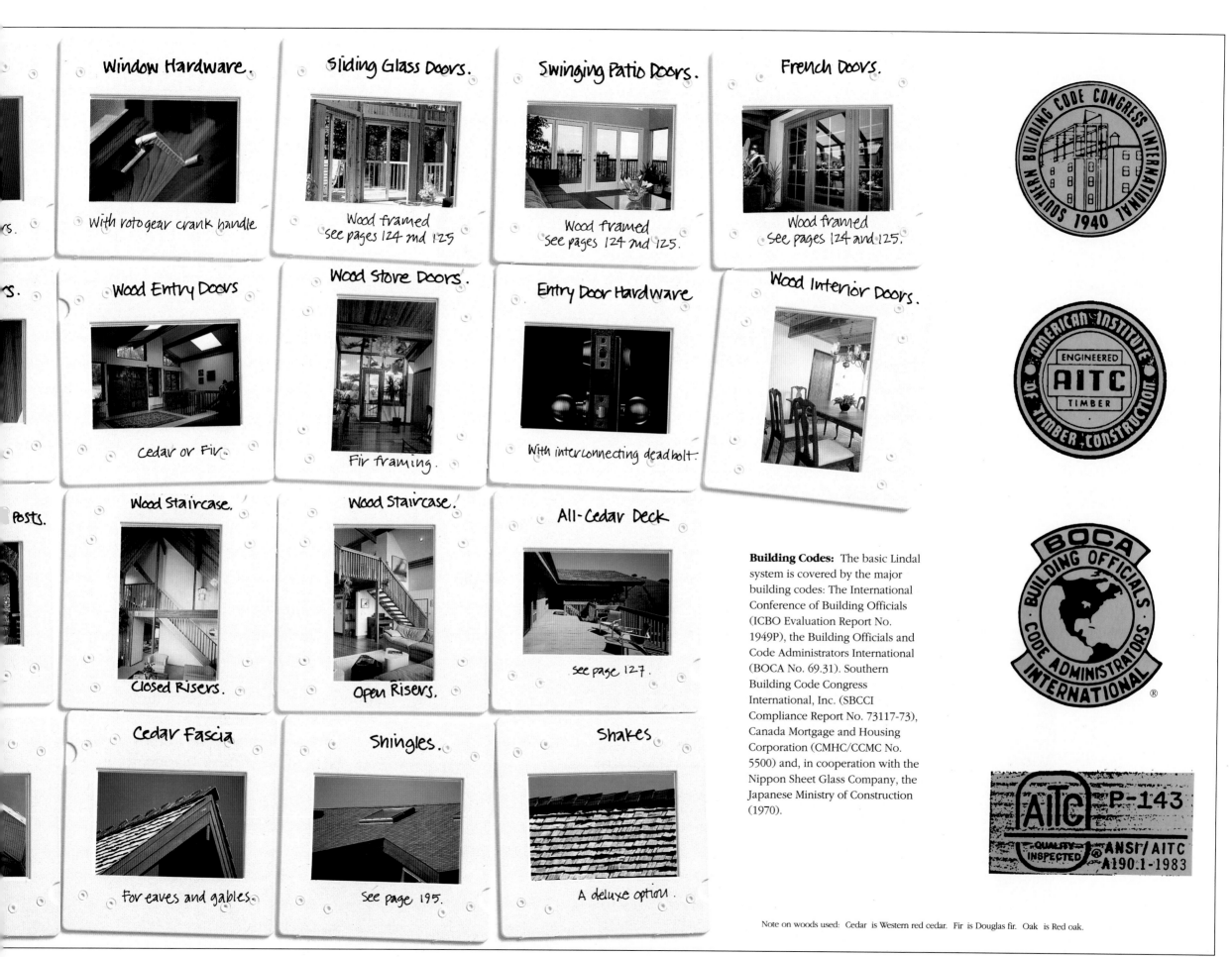

Window Hardware.
With roto gear crank handle.

Sliding Glass Doors.
Wood framed
See pages 124 and 125

Swinging Patio Doors.
Wood framed
See pages 124 and 125.

French Doors.
Wood framed
See pages 124 and 125.

Wood Entry Doors
Cedar or Fir.

Wood Stove Doors.
Fir framing.

Entry Door Hardware
With interconnecting dead bolt.

Wood Interior Doors.

Wood Staircase.
Closed Risers.

Wood Staircase.
Open Risers.

All-Cedar Deck
See page 127.

Building Codes: The basic Lindal system is covered by the major building codes: The International Conference of Building Officials (ICBO Evaluation Report No. 1949P), the Building Officials and Code Administrators International (BOCA No. 69.31). Southern Building Code Congress International, Inc. (SBCCI Compliance Report No. 73117-73), Canada Mortgage and Housing Corporation (CMHC/CCMC No. 5500) and, in cooperation with the Nippon Sheet Glass Company, the Japanese Ministry of Construction (1970).

Cedar Fascia
For eaves and gables.

Shingles.
See page 195.

Shakes
A deluxe option.

Note on woods used: Cedar is Western red cedar. Fir is Douglas fir. Oak is Red oak.

Massachusetts

Horizontal Glue-Laminated Roof and Loft Beams at 5'4" on-center. Built of kiln-dried Douglas fir, our architectural grade, glue-laminated beams do not split, warp, twist or check the way solid wood beams can. They're structurally more sound, too.

Washington

Supporting Posts. The roof's weight is supported by sturdy posts and beams, rather than the partitions on which conventional construction depends. In standard Lindal specifications, exposed posts are glue-laminated to prevent splitting and assure a lifetime of good looks.

A Superior Floor System. Lindal's post and beam system starts with a strong, rigid floor that stays that way. By placing our floor beams closer together – 5'4" on-center – and topping them with 2x6 joists that are glued and nailed to a sub-floor of 3/4" tongued and grooved plywood, Lindal floors are virtually immune to the squeaks and bounces of so many floors today.

POST & BEAM

*L*indal's unique building system, developed and proven over more than forty-five years, is based on post and beam architecture – the choice of North American masterbuilders. Unlike today's conventional construction, in which the walls support the roof's weight, the Lindal system uses a strong framework of posts and beams to support the weight of the roof. The beams are placed 5'4" on-center – the Lindal module. When you work with our planning grid and ruler, you'll see that these tools are based on the same modular system.

What does post and beam construction mean to you? It is literally the key to the design flexibility you enjoy in a Lindal home. Since the walls are not structural, they can be moved easily, which makes it easy to customize any Lindal floorplan to meet your most personal wishes – functionally and aesthetically. An added plus: post and beam opens up the interior, allowing for expansive spaces that make the most of cedar's natural beauty. What's more, the timeless appeal of post and beam is enhanced by Lindal's use of top grade kiln-dried framing materials throughout your home; even those hidden from view are top quality.

Our post and beam homes come in four major styles: Lindal, Clapboard, Round Log and Justus. Before we describe what makes each one unique, here's a look at their common strengths.

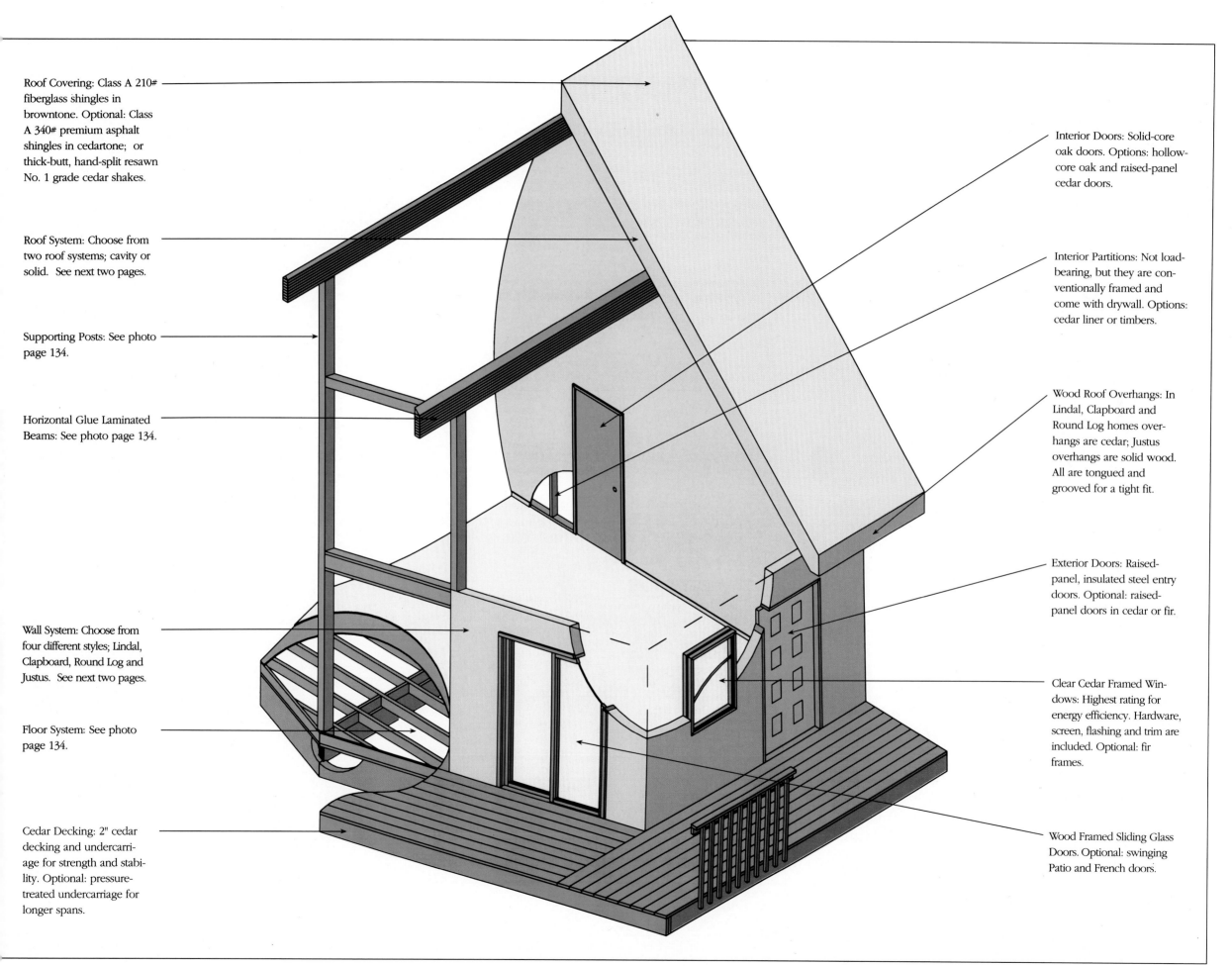

Roof Covering: Class A 210# fiberglass shingles in browntone. Optional: Class A 340# premium asphalt shingles in cedartone; or thick-butt, hand-split resawn No. 1 grade cedar shakes.

Roof System: Choose from two roof systems; cavity or solid. See next two pages.

Supporting Posts: See photo page 134.

Horizontal Glue Laminated Beams: See photo page 134.

Wall System: Choose from four different styles; Lindal, Clapboard, Round Log and Justus. See next two pages.

Floor System: See photo page 134.

Cedar Decking: 2" cedar decking and undercarriage for strength and stability. Optional: pressure-treated undercarriage for longer spans.

Interior Doors: Solid-core oak doors. Options: hollow-core oak and raised-panel cedar doors.

Interior Partitions: Not load-bearing, but they are conventionally framed and come with drywall. Options: cedar liner or timbers.

Wood Roof Overhangs: In Lindal, Clapboard and Round Log homes overhangs are cedar; Justus overhangs are solid wood. All are tongued and grooved for a tight fit.

Exterior Doors: Raised-panel, insulated steel entry doors. Optional: raised-panel doors in cedar or fir.

Clear Cedar Framed Windows: Highest rating for energy efficiency. Hardware, screen, flashing and trim are included. Optional: fir frames.

Wood Framed Sliding Glass Doors. Optional: swinging Patio and French doors.

OUR FOUR MAJOR HOME STYLES

Lindal, Clapboard and Round Log

One might think of this trio of home styles as a family; each shares structural similarities, but has its own unique personality on the exterior.

All three are double-wall frame homes (also known as "double-wall" or "cavity construction"). The exterior cladding is kiln-dried, select grade Western red cedar. The interior wall is drywall. Between the inner and outer walls of all three styles you'll find the same fiberglass insulation – with the same high R values.

A complementary cavity roof system tops these three home styles. Filled with fiberglass batt insulation, the roof has a top R value rating. Drywall lines the roof interior; premium cedar ceiling liner is a popular option.

Justus

An elegant and energy-wise descendant of the historic log cabin, the Justus home pioneers a new standard in the strength and sophistication of solid wood homes.

The integrity and insulation of the Justus wall are not compromised by ill-fitting round logs. Justus walls are made of four-inch thick, finely finished solid cedar timbers, edged with double tongues and grooves to lock together for a zero tolerance fit. That's why we call it our Energy Lock Wall. At intersecting walls, the optimum woodwork-

LINDAL BUILDING SYSTEM

Double-wall frame construction

Lindal (Double-wall frame construction)
The outer walls can be Lindal vertical cedar siding, Clapboard horizontal cedar siding or Round Log horizontal cedar rounds. Inside, walls and ceiling are drywall; opt for as much cedar inside as you wish.

Round Log
Eight-inch log rounds on the exterior.

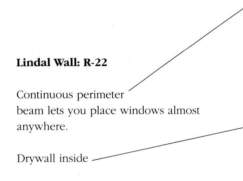

Clapboard
Horizontal cedar siding on the exterior.

Lindal Roof: R-33

Fiberglass shingles

Bulding paper

1/2" plywood sheathing

9" batts of fiberglass insulation

2x12 rafters; 24" on-center

Continuous vapor barrier

5/8" drywall lines the interior

Lindal Wall: R-22

Continuous perimeter beam lets you place windows almost anywhere.

Drywall inside

2x2 horizontal furring; 24" on-center

Continuous vapor barrier

Second (interior) layer of insulation

2x4 vertical studs; 24" on-center

First (exterior) layer of insulation

1/2" plywood sheathing

Tyvek®housewrap

Cedar siding: Lindal, Clapboard or Round Log

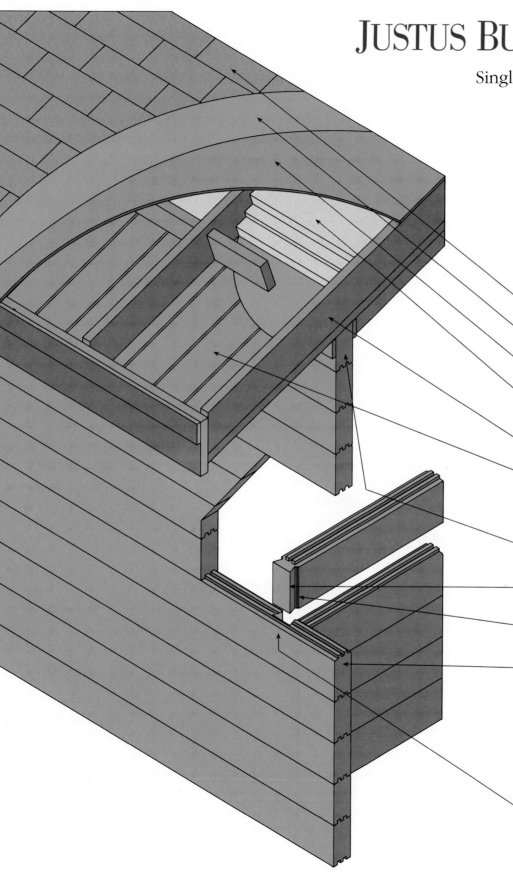

JUSTUS BUILDING SYSTEM

Single-wall solid wood construction

Justus (Single-wall solid wood construction) – Exterior walls are constructed of solid, four-inch thick cedar timbers positioned horizontally. Interior partition walls are drywall, with cedar timbers optional.

Justus Roof: R-38

Fiberglass shingles

Building paper

1/2" plywood sheathing

4" of solid foam thermal insulation plus 1" of styrene

2x8 rafters; 24" on-center

2" thick, double tongue and groove solid-wood planks

Justus: Energy Lock Wall

Top timber acts as continuous perimeter beam for flexible placement of windows.

Diagonal cuts wedge the timbers tight to prevent shearing, racking and lateral pressure.

Second diagonal jog provides added defense against weather.

Solid 4" thick double tongue and groove cedar timbers.

Note: Timber does not penetrate through to the outside at wall intersections – for superior protection against weather and air infiltration.

ing joint – the dovetail – is used to prevent movement and slippage.

Our kiln-dried cedar adds to the airtight fit and energy efficiency of Justus homes. While many manufacturers use green wood, which results in shrinking, twisting and warping, Justus cedar timbers are dimensionally stable from the moment they come out of our kilns. That allows us to machine each timber to an exacting fit. Under normal building conditions, no straps, cables, tie rods or bolts are needed to hold together a Justus home – unlike many manufactured log homes.

These major improvements over common log construction eliminate the energy and settling problems typical of log homes. In fact, the Justus fit is so tight that no caulking is required between the timbers. And, because the timbers are factory precut for each home and individually labeled for course (or layer) and location, construction is fast and easy.

All of the cedar in a Justus home makes it much heavier than a conventional home. The thermal mass created by the four-inch thick cedar timbers provides the equivalent thermal performance of R values used in more conventional housing.

The Justus home comes standard with a two-inch thick, double tongued and grooved plank roof, insulated with rigid foam for excellent energy efficiency and a high R value rating.

Note: You can substitute the Lindal cavity roof on the Justus home – just as you can put a solid plank roof on a Lindal, Clapboard or Round Log home.

STRONG, EFFICIENT BUILDING SYSTEMS— FROM THE GROUND UP

A master-crafted Lindal home not only looks good; it works hard, too. One important aspect is its energy efficiency: your home's ability to keep you comfortably warm in the winter, cool in the summer – and to do it cost-effectively.

Cutting Energy Costs

There's little doubt that energy prices will go up. But when you're building a new Lindal, you have a rare opportunity to exercise significant control over your future fuel costs.

Lindal, Clapboard and Round Log homes depend on insulation for energy efficiency; the Justus specification combines insulation with the thermal mass of its solid wood construction. Both work effectively because they work within the framework of a tightly constructed house, locking out costly air infiltration – and locking in expensive energy.

Build In Energy-Saving Features Now

Most people are surprised to discover that even small crevices around electrical outlets, vents and ducts can result in substantial heat losses and air infiltration. The same expensive losses can occur around the doors and windows. Even insulation and weather stripping must be properly installed to do their job. It's easy to install critical energy features in your home during the building stage; it's

difficult, expensive and sometimes impossible to add them later. Keep in mind, too, that the dollars you invest to reduce fuel bills now will work even harder for you in the future, when energy is bound to be more expensive. Your energy-efficient home will also pay off further down the line; it will be more saleable and more valuable than the typical under-insulated house on the market.

Energy-saving features that keep a house warm in the winter work equally well to keep it cool in the summer. In extremely hot climates, insulation dramatically reduces the size of the air conditioning system needed and can shorten the length – and expense – of your cooling season.

Fit Your Lindal to Your Climate

Your insulation requirements will depend on the climate in which you build. Also, states and provinces have varying code requirements. By carefully selecting from our wide variety of energy options, you can fine-tune your Lindal home to these local considerations.

Measuring Insulation Value

In this chapter, you'll be reading a lot about R values. The insulating ability of a home's components – whether it be the roof, walls, floors or doors – is measured in terms of R value. The R value indicates a material's ability to resist the passage of heat. The higher the R value, the better the material resists its natural inclination to let heat escape out the house.

R and U Values

All R and U values stated are based on the American Society of Heating, Refrigeration and Air Conditioning Engineers, Inc. (ASHRAE) Handbook of Fundamentals and have been independently substantiated by Siefert & Forbes, Engineers, Tacoma, Washington.

R Values in Canada are calculated differently and vary slightly from the U.S. values given here.

LINDAL FLOORS: STANDING UP TO A LIFETIME OF USE

*O*ver the years, your floor will get more wear and tear than just about any other part of your home. You want it to withstand decades of everyday use without getting loose or squeaky in the process.

That's why Lindal's strengths begin with our floor system – and why, like many Lindal features, our floors are a departure from today's common construction. The standard Lindal floor is designed for lasting strength and rigidity in three important ways:

1) Our floor beams are placed closer together. Lindal floors consist of two 2x10 floor beams 5'4" on-center, over which 2x6 floor joists are positioned 16" on-center.

Spacing the floor beams closer together takes some of the weight off of the floor joists, which alleviates the problems of irritating bouncing and squeaking underfoot.

2) We block our 2x6 joists rather than cross-bridging them. That keeps them more tightly in place than common cross-bridging, contributing to a stable subfloor – and eliminating yet another of the underlying reasons for bounces and squeaks.

Lindal Permanent Wood Foundations

Lindal supplies well-insulated permanent wood foundations that are as strong and long-lasting as any type of foundation money can buy – without the inherent risks of a concrete foundation, such as cracking and water leakage. Other advantages to our wood foundations: they're easily brought into building sites that are difficult for concrete trucks to reach; many of our foundations are barged into sites where road access is difficult. What's more, the same crew that builds your house can build your permanent wood foundation, which tends to save money and add continuity to quality control.

Designed to be buried into the soil, the bottom bearing plate, studs, posts and plywood of our foundations are made of pressure-treated wood, wrapped in a poly moisture barrier and held together with rust-resistant pressure treated lumber nails. These materials resist rot and ravages of time and are backed by a 30-year guarantee from the manufacturer – a rarity in construction materials today.

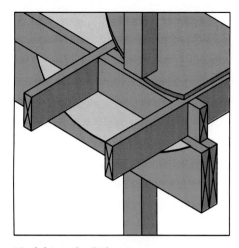

**Lindal Standard Floor
R-12**

3) We use 3/4" underlay-grade tongue and groove plywood to create a strong interlocked subfloor. Furthermore, the interlocking edges of the plywood are glued together, and the entire plywood subfloor is glued and nailed with ring shank nails for more holding power over the floor joists.

Finally, a layer of reflective foil insulation is applied under the floor. This APA glued subfloor has an R value of 12 – a vast improvement in energy efficiency over most conventional floor systems today.

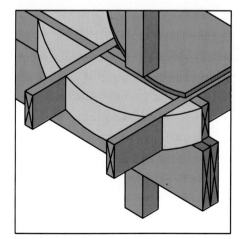

**Lindal Polar Floor
R-21**

Floor Insulation – How Much?

The amount of floor insulation required varies greatly from home to home, depending on your climate and what you're building over: a crawl space, basement, post and pier, piling, concrete slab or permanent wood foundation.

If you plan to build on a crawl space and your climate is especially severe, we recommend the optional Lindal Polar floor, which packs six inches of insulation between the floors joists and boosts the R value to 21.

However, if you're building on a full basement, you probably won't need the extra insulation. Instead, you'll be better off using one inch or more of rigid insulation to line the exterior of your basement walls, bringing them up to an R value of 5 or more. Or consider a permanent wood foundation with R values of 20 to 30.

LINDAL WALLS: KEEPING THE ENERGY IN

Whether you are building a double-wall home (Lindal, Clapboard or Round Log) or a single-wall, solid wood Justus, you can count on a wall system of superior integrity, energy efficiency and noise insulation.

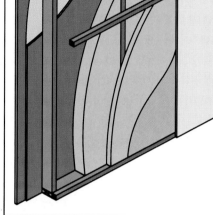

**Lindal Standard Wall R-22
Lindal Polar Wall R-28**

Lindal, Clapboard and Round Log Walls

The standard Lindal wall used in Lindal, Clapboard and Round Log homes is more energy-efficient than conventional walls.

From the outside in, here's what makes up this 7 1/2" thick wall: First, the exterior wall of your choice (Lindal vertical siding, Clapboard horizontal siding or Round Log horizontal siding) applied over Tyvek® housewrap and one-half-inch plywood. Most important,

---

A Warm, Snug Justus in Maine

"After babysitting a Lindal cedar home for friends ten years ago," writes Lindal homeowner Paul Carr, "I decided I was going to have one for myself."

Paul and his wife, Lynne, built their Justus style "dream house" in Kennebunkport, Maine. "The brilliant sunlight streaming in through the Prow windows, the open, airy spaciousness of the soaring cathedral ceilings, and the warm, comfortable feeling of solid cedar are more than we had dreamed it would be."

What's more, Paul writes, their Justus home is an energy-efficient dream even during Maine's most severe weather. "Our Justus costs next to nothing to heat, even in the coldest months. I'd recommend a Justus home to anyone."

---

you'll find two layers of insulation. One overlaps the framing itself, eliminating the customary gaps in insulation. Inside, on the warm side, the interior layer of insulation is sealed with a vapor barrier.

This unique cavity wall system adds up to exceptional wall insulation – at an R value of 22.

Double-Wall Insulation: How Much?

Since wall insulation provides a major improvement in R values at a small price, you'll want to add as much as you can.

If you live in an extremely cold region, we recommend the optional Lindal Polar Wall. This 9" thick wall has an R value of 28.

Housewrap and Continuous Vapor Barrier

Tyvek® housewrap is a lightweight sheeting of high-density woven polyethylene fibers that wraps the house between the cedar siding and the exterior, or cold, side of the insulation. It reduces heat loss by as much as 33 percent. And, because it lines the exterior, openings in the wrap are minimized, cutting air infiltration by up to 90 percent. Tyvek® lets moisture pass through, so there is no danger of condensation in the walls. And it's good for the life of your home.

Lindal's continuous vapor barrier of heavy polyethylene sheeting is applied to the interior, or warm, side of the insulation. It, too, acts as an air infiltration barrier. It also prevents moisture vapor from entering the insulation, which can destroy its insulating value and cause rot.

Justus: Energy Lock Walls in Solid Wood

Solid Justus walls do not rely on conventional fiberglass insulation for their energy efficiency. Instead, they rely on extremely tight construction and the thermal mass of their energy-retaining cedar timbers.

The solid four-inch thick cedar timbers that make up Justus walls have double tongued and grooved edges, which lock together for a zero tolerance fit. Where walls intersect, we use the strongest woodworking joint of all: the dovetail. Together, this combination of tongue and groove and the dovetail form the Justus' Energy Lock system for airtight construction.

The solid cedar in a Justus home makes it much heavier than a conventional home. All that mass provides the equivalent thermal performance of R values in more conventional homes.

Thermal Mass: Another Measure of Energy Efficiency

Justus homeowners rave about the warm, even temperatures they can maintain at low energy costs.

Here's how thermal mass works: During the winter, the heavy, solid wood walls absorb and store heat from the sun and/or interior heating systems, radiating that heat slowly and evenly back into the house as the temperature drops during colder hours. By holding onto costly heated air and releasing it back into the house as the temperature drops, energy consumption is reduced. During the summer, the solid walls absorb the heat of the day and radiate it slowly during the cooler, night hours. This natural process of storing and time-releasing heat is

known as thermal lag – and it's the reason why Justus homes are cozy and warm in the winter, cool in the summer.

The energy-saving impact of thermal mass has been widely studied – and just as widely acknowledged. In the early 1980's a study was conducted by the National Bureau of Standards (NBS) for the Department of Health and Urban Development (HUD) and the Department of Energy (DOE) to determine the effects of thermal mass on a building's energy consumption. The study conclusively proved that solid wood structures are highly energy efficient. So while the principle behind a Justus' energy efficiency is different from our double-wall specifications, the result is the same: a comfortable, even-temperatured living environment 365 days a year – and lower energy costs.

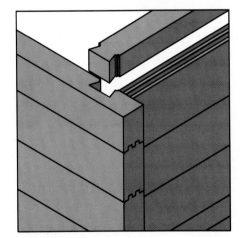

Justus Standard Wall

Lindal Tropical Specifications

Lindal offers three specifications designed to make you at home in any tropical climate. All are designed to eliminate unnecessary insulation, make the most of ventilation, and take advantage of cedar's natural resistance to insects and decay.

Double-Wall Tropical. A double-wall frame home with vertical tongue and groove siding of 1"x6" resawn cedar on the outside; drywall on the inside – and no unneeded insulation or furring in between.

Single-Wall Tropical. A single-wall solid wood home with vertical tongue and groove siding of 2"x8" smooth-face, pre-cut cedar planks.

Tropical Justus. The single-wall, solid wood of our regular Justus specification is modified for tropical comfort and beauty; 3x8 solid cedar timbers, double tongued-and-grooved, are laid horizontally.

Note: For more information on Lindal's tropical home designs and specifications, see our Tropical planbook and specification sheet.

LINDAL ROOFS: TOPS IN QUALITY, ENERGY EFFICIENCY

*M*ore heat escapes out of the roof than the walls – which is why your roof needs more insulation. As a general rule, buy as much as you can afford. The price of additional insulation is not costly when you're building your home – but it's expensive to add later.

Lindal Standard Roof

Lindal's cavity roof system is standard with Lindal, Clapboard and Round Log homes. This R-33 roof is recommended for most climates. Its 2x12 rafters are filled with 9" batts of fiberglass insulation and topped

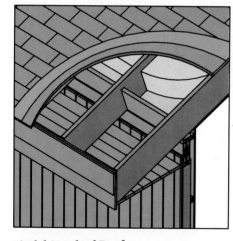

Lindal Standard Roof
Solid 2x12 Rafters. R-33

by 1/2" plywood sheathing. A continuous vapor barrier is placed against the underside of the rafters on the warm side. Drywall lines the inside ceiling. Shingles top the roof's exterior. All roof beams are horizontal glue-laminated, sized for your home's individual roof load.

Because 2x12 rafters are the deepest solid rafters available, it's physically impossible to increase the roof insulation beyond 9" using a rafter system. To overcome this limitation, we have developed our own Polar Cap line of roofs – with rafters that can be expanded from a standard depth of 16" all the way to 22".

Polar I – For Warm, Humid Climates

Our Polar Cap I roof is ideal for warm, humid areas where summer

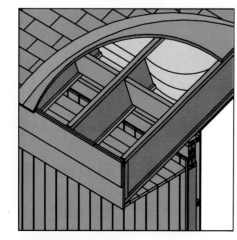

Lindal Polar Cap I and II Roofs
Trus Joist I-Beam Roof
16" Cavity. R-33 to R-41

cooling and humidity are problems. This R-33 roof, with its 9" of insulation, is ideal for minimizing air conditioning losses. The roof's generous air space and vents virtually eliminate the possibility of heavy condensation damaging the insulation – a common problem in warm, humid parts of the world.

Polar II – For Cold Climates

For energy-efficient comfort in cold climates, move up to our Polar Cap II. It uses two 6" layers of R-19 insulation – for a total roof R value of 41.

Polar III – For Extreme Cold

For frigid climates like Alaska's, we recommend Lindal's unique Polar Cap III roof. Its 22" cavity can hold two 9" layers of R-30 insulation for an amazing total roof R value of 63.

Lindal Polar Cap III Roof
Expandable E Rafter Roof
22" Cavity. Up to R-63

Justus Standard Roof

The solid wood Justus roof is a 2" thick, double tongued and grooved plank roof. Between the 2x8 rafters, which sit over the roof decking, are 4" of solid foam thermal insulation plus 1" of styrene. The roof is then topped with 1/2"

Justus Standard Roof
Solid Plank Roof. R-38

plywood and building paper and covered with fiberglass shingles. Inside, you have all the natural warmth and beauty of a wood ceiling. This standard Justus roof has a highly efficient R value of 38 – offering a solid alternative to the Lindal cavity roof. Like the cavity roof, it can be used interchangeably; you can opt for it to top a Lindal, Clapboard or Round Log home, if you wish.

Roof Ventilation

Ventilation is critical to preventing rot and mold in your roof. That's why all Lindal roof systems include a continuous soffit vent and a ridge vent that runs the entire length of the soffit and ridge. Together, they create a natural, continuous flow of air that keeps the roof insulation and framing dry, preventing the moisture that can cause rot and mold. This system also vents out unwanted heat in summer which can save on air conditioning.

Lindal Polar Options: Energy-Wise Comfort for Extreme Climates

Lindal's Polar Options are designed for the ultimate in comfort and energy efficiency in extreme climates. While most are geared to especially cold climates, we recommend our Polar Cap I roof for warm, humid climates such as Florida.

Lindal Polar Floor – R-21

Ideal for building over a crawl space in extremely cold climates.

Lindal Polar Wall – R-28

A 9" thick insulated wall for extremely cold regions.

Lindal Polar Cap I Roof – R-33

For warm, humid climates. Minimizes air conditioning losses and condensation.

Lindal Polar Cap II Roof – R-41

Added insulation for cold climates.

Lindal Polar Cap III Roof – Up to R-63

Two 9" layers of insulation; for severely cold regions such as Alaska.

LINDAL WINDOWS: CLEAR ADVANTAGES

As you plan the windows in your home, you'll want to keep in mind that they can be a major source of energy loss. Windows have very little insulating value compared to the walls and roof of your home. Even so, there's good news about energy-efficient glazing options.

Lindal's standard wood framed double-glazed windows are a great place to start – and you can add state-of-the-art coatings, films and fillings to them for even higher energy efficiency.

Our Standard: The Highest

The exclusive Lindal-manufactured window has earned the highest industry rating available, "Class A Improved," which minimizes air and water infiltration under maximum testing conditions.

All Lindal windows are framed in wood instead of aluminum. Wood has a natural insulating capacity, whereas aluminum actually conducts heat out – 2,000 times faster than wood. What's more, wood does not "sweat" like aluminum. It looks better, too. The window frame is fitted with two strips of continuous weather stripping – much like a refrigerator – to stop air infiltration and reduce drafts.

Lindal's double-glazed windows have an R value of 2 and a U value of .50 – more than twice the energy-efficiency of single panes. (We don't recommend single-pane windows except for use in the tropics and in uninsulated garages.)

If you want to go beyond Lindal's standard double-glazed windows, we recommend Low "E" glass, Heat Mirror® glass or Argon Low "E" glass. In each case you are adding the latest technology in coatings, films and fillings – all of which increase energy efficiency and reduce cold spots, drafts, condensation and fabric fading.

Optional Low "E" Glass

This special transparent coating, applied to ordinary double-glazed glass, lets sunlight enter your home and keeps it inside by reflecting it back into the room. In hot weather, Low "E" also reduces heat gain. With a U value of .42, Low "E" glass is much more effective than standard double-glazed windows.

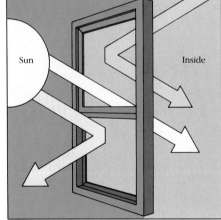

The secret of optional, energy-smart window glazings: they transmit short wavelength visible rays and reflect long wavelength heat rays. So they keep your home warmer in cool weather; cooler in warm weather.

Optional Heat Mirror® Glass

Lindal windows can incorporate Heat Mirror® technology, in which a layer of transparent polyester film, with a heat-

---

Making Sense of R and U Values

While R values are a standard measure of insulation for every other part of the home, the window industry talks in terms of U values. So you'll want to speak the language – or at least understand it – to get the most energy-efficient windows for your home.

Just to make things a bit more complicated for the layperson, R and U values are inversely related: the higher the R value, the lower the corresponding U value. In fact, the R value is derived from the U value: the actual formula is U=1/R. (For example, a U value of .25 would equal an R value of 4.) But for your purposes, what you really need to know is this: the higher the R value, the higher the energy efficiency. The lower the U value, the higher the energy efficiency. The bottom line: look for low U values and high R values in your windows.

Profile of Our "Class A" Improved Window

Quality Frame Material: Lindal's standard window frames are clear, fine-grained Western red cedar with no knots, no voids. Fir frames, with a thinner profile, are optional.

Fiberglass Screens. A Lindal standard with all opening windows.

Standard Double-Glazed Glass. Reduces heat loss, drafts, noise and condensation build-up.

Mortise and Tenon Joints. Used for the wood joints of our casement and awning windows for an airtight, watertight fit.

Quality Window Hardware. Smooth, effortless operation.

A Variety of Glazing Options. From double glazed to state-of-the-art Argon Low "E". Also clear, obscure and tinted.

Two Strips of Continuous Weather Stripping. For airtight, watertight energy efficiency.

reflective coating – like Low E glass – is stretched tight and sealed between the two panes of glass. This creates two air spaces which add to Heat Mirror®'s energy efficiency. Tests conducted by the National Certified Testing Laboratories certify that heat Mirror® is 90 percent higher in R value than double glass, and 58 percent better than Low E glass.

A Smart Option:
Argon Low E Glass

We're proud to pass the latest advance in window technology on to you: Lindal's optional Argon Low E Window Systems have a U value of .29 – nearly as low as Heat Mirror® – at a substantially lower price.

This option replaces the air between the double panes with Argon gas, which has been used successfully and safely in Europe for more than 20 years. Argon gas is inert, colorless, non-corroding and non-reactive – with a lower heat conductance than plain air. As a result, it generates a better U value and increases comfort.

Designing with Windows

Be practical in choosing the number and type of windows for your home. Look for ways to reduce their number and size – and place them where you'll really enjoy them.

Maximize glass on the south side for passive solar gain; minimize glass on the north side to protect against cold winter winds.

Lindal Windows		
Value	R	U
1. Double Glazed	2	.50*
2. Low E	2.4	.42
3. Argon Low E	3.4	.29
4. Heat Mirror®	3.8	.26
• Clear		
• Obscure		
• Tinted		

Note: The frames add significantly to the total U value. For example, our cedar framed, double glazed window has been tested and certified at U.37 – 26% better than the U value for the glass alone.

LINDAL DOORS: THEY MAKE A GRAND ENTRANCE

*T*he door to your home is a very personal choice – and it's one of the places where people sometimes sacrifice energy efficiency for looks.

Exterior Doors

For the ultimate in energy efficiency – and security – nothing beats a metal door system. Lindal provides a top-of-the-line insulated entry door system, complete with weather stripping for a tight fit. This 24-gauge galvanized steel door, with polyurethane core, has an R value of 14, compared to a regular wood door's R value of 2.

Lindal's optional raised-panel wood entry doors are available in Western red cedar or Douglas fir.

Glass Doors: Sliding Glass, Patio, French and Store Doors

While glass doors have little insulating value, they are popular for the large expanses of glass and light they provide.

We carry only wood framed glass doors for the same reason we carry only wood framed windows: Wood is a natural insulator, unlike aluminum, which actually conducts heat *out.* Aluminum is also prone to condensation and drafts.

To conserve energy, glass doors should be weather stripped, with double glazing and a thermal break system that virtually eliminates sweating and frost build-up. Lindal wood framed glass doors come with all of these features. They are double-glazed, with weather stripping, and the wood itself provides the best thermal break available.

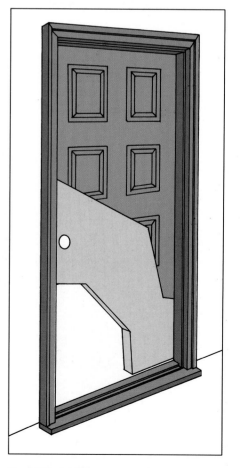

Lindal Entry Door
Galvanized Steel; R-14

GOING SOLAR

*O*ne solution to the high cost of energy is harnessing free energy from the sun. By building an energy-efficient, passive solar home, you can reduce fuel needs dramatically – as much as 40 to 90 percent.

Site orientation is the easiest and least expensive way to take advantage of solar energy. By facing a wall of glass to the south, your home becomes a passive solar collector that stores and distributes the sun's energy without requiring expensive mechanical equipment such as active solar collectors. Remember, passive solar techniques work equally well to keep your home cool and cut down on air conditioning costs.

"Our utility bill is 40 percent less than our neighbor's with a comparable size house. Would we recommend a Lindal cedar home? Without reservations!"
Dr. Guy and Colette Ulrich, Arkansas

Arkansas

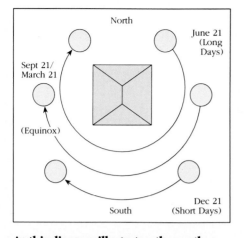

As this diagram illustrates, the south face gets the most sun; the north gets the least.

South-facing glass is essential to collect the sun's free heat. The roof overhang is carefully sized to admit the low winter sun's heat, yet blocks the high summer sun. Deciduous trees contribute to summer shade; come winter, they let sunshine stream in.

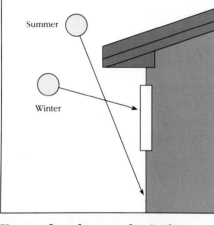

Your roof overhang can be sized to block the high summer sun, but let in the low winter sun.

Once glass has trapped the sun's energy, interior mass is needed to store it. Here the mass is contained in a thermal storage wall, which is faced with local field stone.

PICK UP SOME IDEAS FROM THIS SOLAR CLASSIC.

*T*his custom Lindal home incorporates just about every solar technique we've mentioned – and more. From its thermal storage wall to the active solar collectors on its roof, it's a classic of solar-wise design. Yet it was constructed for about the same price as a similarly insulated, conventionally heated house in the same area. What you see here dispels two myths about going solar: one, that a solar home is more expensive to build; and, two, that, by nature, it is unattractive or bizarre in appearance.

As this photograph shows, a well-designed solar home is an aesthetic asset to its surroundings. Unlike its neighbors, however, most of this home's heating requirements are fueled by the sun. A wood-burning stove provides the rest. Simply by using its passive solar techniques, you, too, can enjoy savings of 40 to 90 percent on your energy bills.

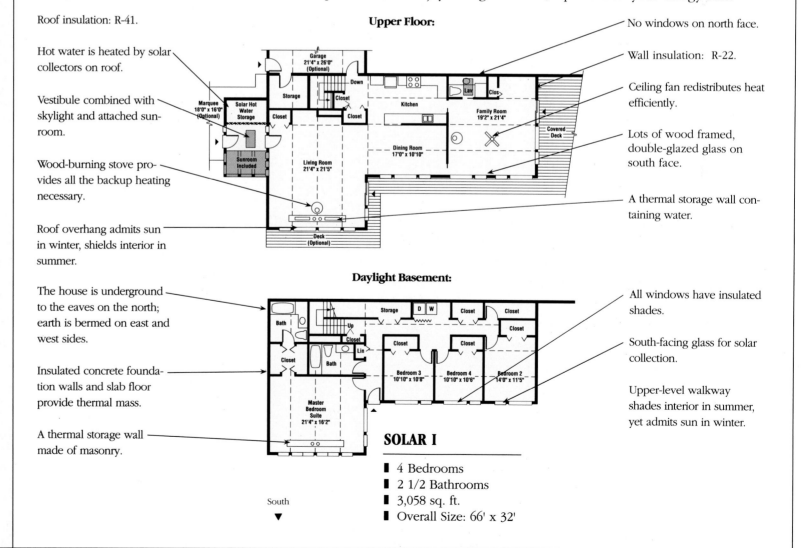

Roof insulation: R-41.

Hot water is heated by solar collectors on roof.

Vestibule combined with skylight and attached sunroom.

Wood-burning stove provides all the backup heating necessary.

Roof overhang admits sun in winter, shields interior in summer.

The house is underground to the eaves on the north; earth is bermed on east and west sides.

Insulated concrete foundation walls and slab floor provide thermal mass.

A thermal storage wall made of masonry.

Upper Floor:

No windows on north face.

Wall insulation: R-22.

Ceiling fan redistributes heat efficiently.

Lots of wood framed, double-glazed glass on south face.

A thermal storage wall containing water.

Daylight Basement:

All windows have insulated shades.

South-facing glass for solar collection.

Upper-level walkway shades interior in summer, yet admits sun in winter.

SOLAR I

■ 4 Bedrooms
■ 2 1/2 Bathrooms
■ 3,058 sq. ft.
■ Overall Size: 66' x 32'

CHAPTER NINE

PLANS

Before you put your ideas down on paper, browse through our design library to take advantage of decades of design and home building experience. Each plan is subject to change. Your change. Discover how easily you can add a window—or a wing. Alter the size of a room or the style of the roof. We've supplied all the tools you need to get started: grid paper, a ruler scaled to our plans, and a template for tracing interior fixtures. Once you've traced and modified your floor plan, or drawn your own one-of-a-kind, your local Lindal dealer will do a complimentary feasibility and cost analysis. Or, if you wish, let your dealer turn your wish list into the perfect plan.

HOW TO PERSONALIZE A PLAN

With the scope and diversity of plans in Lindal's design library, you may well find a home that comes close to your dream. All it takes is a little personalizing – and that's easy with a Lindal.

To the right is an example of a plan that's been personalized. This "before" and "after" shows how one family modified a stock plan – the Venice plan, on page 210 – to add needed floor space and create their unique home, the Venice Vista. The "after" plan indicates just a few of the changes you can make to any of the plans in the design library.

Before:

VENICE

- 2 Bedrooms
- 2 Bathrooms
- 1,016 sq.ft
- Overall Size: 21'x34'

After:

VENICE VISTA

- 3 Bedrooms
- 2 Bathrooms
- 1,834 sq. ft
- Overall Size: 60' x 36'

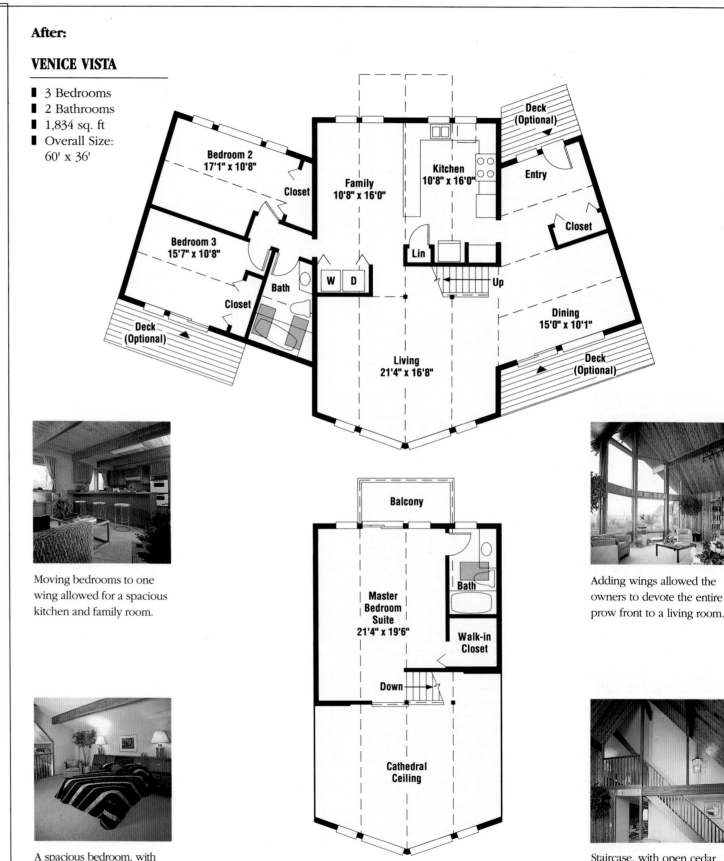

Moving bedrooms to one wing allowed for a spacious kitchen and family room.

A spacious bedroom, with bath en suite, is located upstairs.

Adding wings allowed the owners to devote the entire prow front to a living room.

Staircase, with open cedar balusters and matching cedar toprail, add to the home's expansive feeling.

ANY ROOF ON ANY PLAN

Throughout this book, you have seen many roof styles on many different kinds of homes. Thanks to Lindal's post and beam architecture, you can combine just about any roof with any floorplan. So go ahead – mix and match to your heart's content.

Here's an example of a good mix: with the Venice Vista, pictured opposite, the two-story prow front is topped with a 12/12 roof – like the original Venice plan. But the one-story extensions have a lower, 4/12 roof line. The 12/12 roof could have been a 12/12 or 4/12 or a gambrel. And the 4/12 roof could have been a 12/12 or a gambrel. Of course, these changes in roof line would affect the floorplan layout. Think about it – then have it your way.

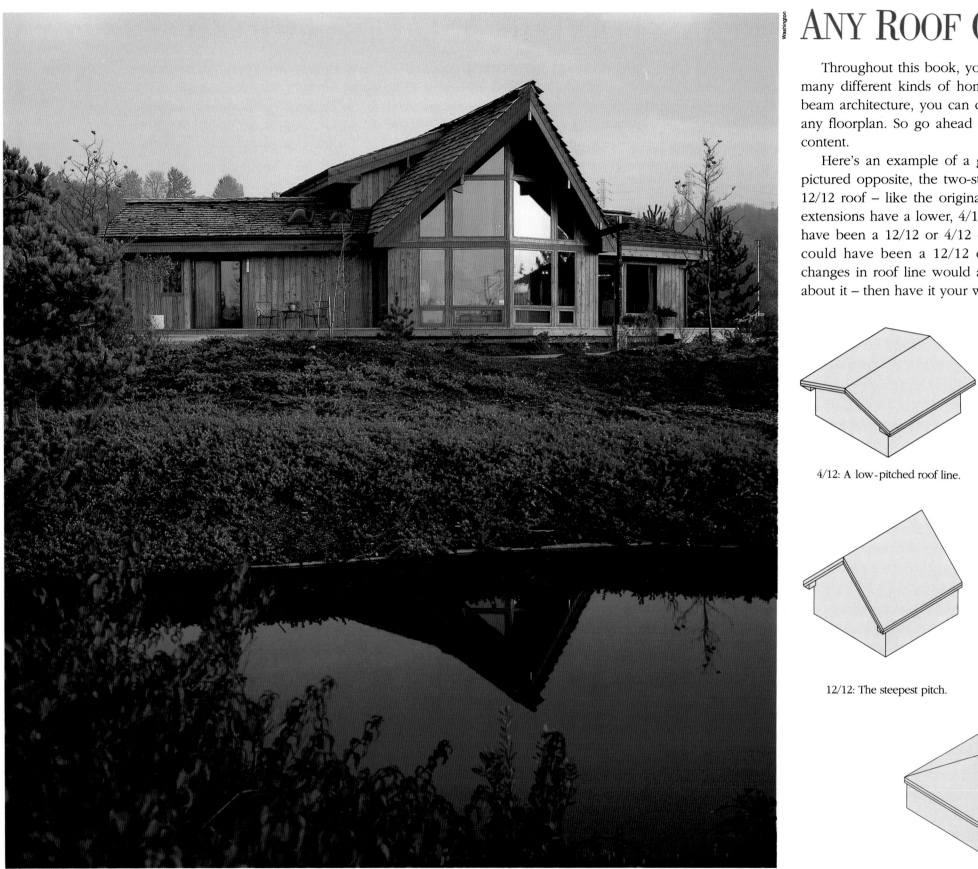

Two 4/12 roof pitch wings flank the 12/12 roof pitch central prow. A 4/12 dormer in the prow increases useable space upstairs.

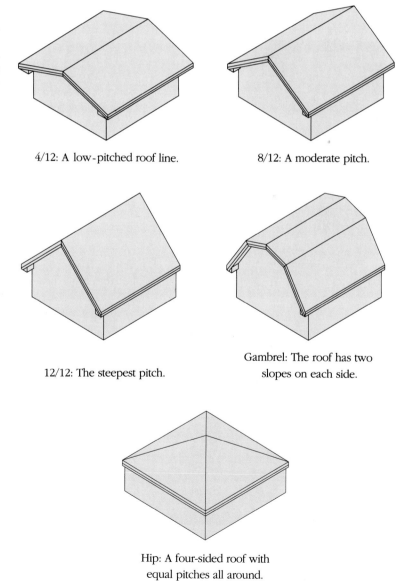

4/12: A low-pitched roof line.

8/12: A moderate pitch.

12/12: The steepest pitch.

Gambrel: The roof has two slopes on each side.

Hip: A four-sided roof with equal pitches all around.

READING THE PLANS: POINTERS AND TERMINOLOGY

*B*efore you head into the plans section, we urge you to take a moment to get acquainted with a few important terms and plan-reading tips:

Optional Features

Plans and photos often show optional features like: garages, decks, porches, skylights, and sunrooms, which can be ordered from Lindal. Plans and photos sometimes show features *not* supplied by Lindal, such as: appliances, fixtures, chimneys, fireplaces, wood stoves, shelving, etc.

Each series offers several variations in floorplans; each variation has its own *model* name.

Within each category of plans, you'll find a *series* of homes sharing common features.

Our homes are divided into four architectural categories: Classic, Contemporary, Traditional and Small Treasures.

This is the *square footage* of the home, including main living area and loft, but excluding optional features such as garages, decks, porches, sunrooms, etc.

This is the overall *dimensions* of the home at its longest and widest points - again, excluding the optional features. Dimensions have been rounded up and down to the closest foot.

This is a handy page number reference to help you find representative photos of homes in the series.

The drawing is a front *elevation* of the home.

The term "Star" indicates that the models in the series have wings.

In two-story homes, wide dark lines in the perimeter of the second floor indicate the outside walls are full height; thinner, lighter lines indicate the roof is sloped, as in the Chalet roof.

NINE / PLANS / CONTEMPORARY HOMES

PRESIDENTIAL PROW STAR SERIES

JACKSON

- 4 Bedrooms
- 2 1/2 Bathrooms
- Master bedroom on second floor
- 2502 sq. ft.
- First Floor: 1470 sq. ft.
- Other Floor: 1032 sq. ft.
- Overall Size: 49' x 38'

Photos of Presidential Prow Stars are shown on pages 72-75.

First Floor

Deck (Optional)

D
W
Utility 11'0" x 10'6"
Dining 16'6" x 16'0"
Family 21'4" x 12'0"
Down
Closet
Up
2-Car Garage 21'4" x 21'4" (Optional)
Kitchen 15'0" x 10'8"
Lav
Entry
Living 21'4" x 21'0"
Deck (Optional)

Class A fiberglass shingles in a browntone to harmonize with the cedar are standard. Consider an opening skylight, too.

Beautiful 1" cedar fascia and soffit trims the eaves.

Second Floor

Balcony
Closet
Closet
Bath
Bath
Bedroom 4 12'6" x 11'6"
Bedroom 2 19'0" x 12'0"
Down
Closet
Master Bedroom Suite 13'6" x 26'8"
Lin
Closet
Bedroom 3 14'0" x 11'6"
Cathedral Ceiling

Thick butt, handsplit and resawn #1 cedar shakes are optional.

Handsome raised panel clear cedar garage doors come in 8 ft., 9 ft. and 16 ft. sizes. They are pre-drilled for easy installation of garage door hardware.

186

© LINDAL CEDAR HOMES

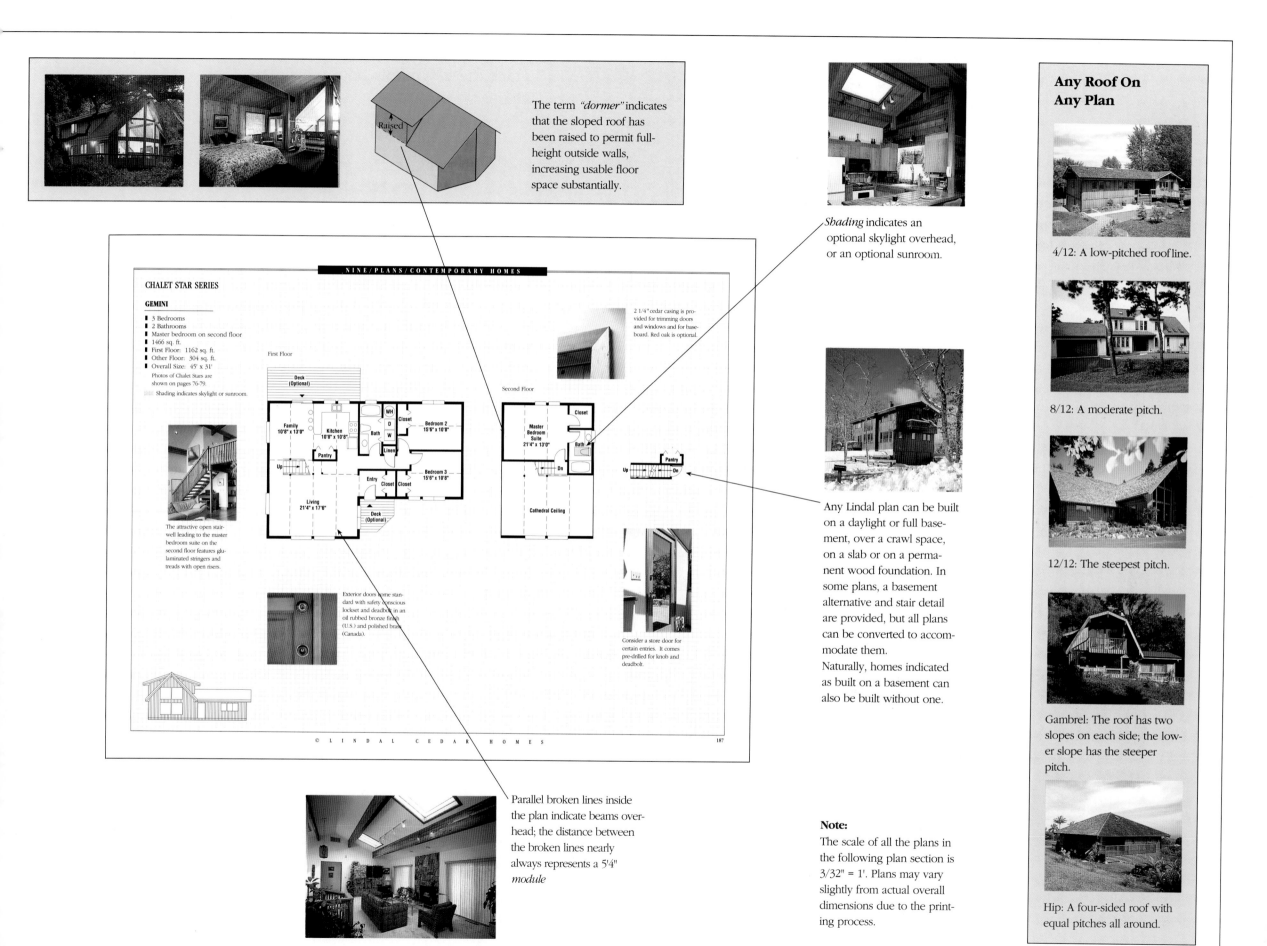

The term *"dormer"* indicates that the sloped roof has been raised to permit full-height outside walls, increasing usable floor space substantially.

Raised

Shading indicates an optional skylight overhead, or an optional sunroom.

2 1/4" cedar casing is provided for trimming doors and windows and for baseboard. Red oak is optional.

NINE/PLANS/CONTEMPORARY HOMES

CHALET STAR SERIES

GEMINI

- 3 Bedrooms
- 2 Bathrooms
- Master bedroom on second floor
- 1466 sq. ft.
- First Floor: 1162 sq. ft.
- Other Floor: 304 sq. ft.
- Overall Size: 45' x 31'

Photos of Chalet Stars are shown on pages 76-79.

Shading indicates skylight or sunroom.

First Floor

Deck (Optional)

Family 10'8" x 13'6"
Kitchen 10'8" x 10'8"
WH D W
Bath
Pantry
Linen
Bedroom 2 15'6" x 10'8"
Closet
Up
Entry
Closet Closet
Bedroom 3 15'6" x 10'8"
Living 21'4" x 17'6"
Deck (Optional)

Second Floor

Master Bedroom Suite 21'4" x 13'0"
Closet
Bath
Dn
Cathedral Ceiling

Up Dn Pantry

The attractive open stairwell leading to the master bedroom suite on the second floor features glu-laminated stringers and treads with open risers.

Exterior doors come standard with safety conscious lockset and deadbolt in an oil rubbed bronze finish (U.S.) and polished brass (Canada).

Consider a store door for certain entries. It comes pre-drilled for knob and deadbolt.

© LINDAL CEDAR HOMES

187

Any Lindal plan can be built on a daylight or full basement, over a crawl space, on a slab or on a permanent wood foundation. In some plans, a basement alternative and stair detail are provided, but all plans can be converted to accommodate them.

Naturally, homes indicated as built on a basement can also be built without one.

Note:

The scale of all the plans in the following plan section is 3/32" = 1'. Plans may vary slightly from actual overall dimensions due to the printing process.

Parallel broken lines inside the plan indicate beams overhead; the distance between the broken lines nearly always represents a 5'4" *module*

Photos of Chalet Stars are shown on pages 76-79.

Any Roof On Any Plan

4/12: A low-pitched roofline.

8/12: A moderate pitch.

12/12: The steepest pitch.

Gambrel: The roof has two slopes on each side; the lower slope has the steeper pitch.

Hip: A four-sided roof with equal pitches all around.

PLANNING AIDS

*Y*ou can do a lot of planning with just two primary tools: the grid paper, opposite, and the ruler included. Remove the grid paper (it's perforated) and use it to trace the floorplan that comes closest to your ideal home – or to draw your own one-of-a-kind. The ruler, with its 3/32" = 1' scale and its module marks, makes its quick and easy to approximate sizes.

Now plan the arrangement of your interior spaces. This can be tricky, since fixtures, appliances and furniture often take up more room than we expect. To help you visualize space requirements, trace the handy symbols, opposite, onto your plan. If you're ambitious, you can even add your windows.

These simple, but smart planning aids make it easy and fun to design your own home. Especially when you use a pencil – with an eraser. You can change your mind dozens of time and erase any mistakes.

If you need free help, don't hesitate to call your local Lindal dealer. You'll find your dealer is trained and ready to offer design advice – as well as a free feasibility and cost analysis of your plan.

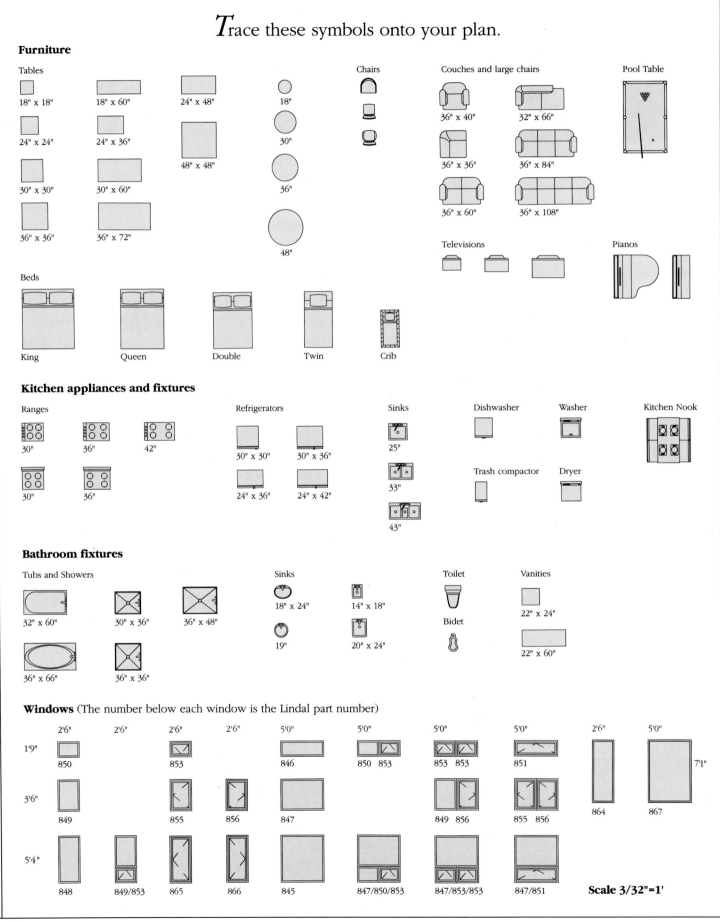

*T*race these symbols onto your plan.

Furniture

Tables: 18" x 18", 18" x 60", 24" x 48", 24" x 24", 24" x 36", 48" x 48", 30" x 30", 30" x 60", 36" x 36", 36" x 72"

Chairs: 18", 30", 36", 48"

Couches and large chairs: 36" x 40", 32" x 66", 36" x 36", 36" x 84", 36" x 60", 36" x 108"

Pool Table

Televisions

Pianos

Beds: King, Queen, Double, Twin, Crib

Kitchen appliances and fixtures

Ranges: 30", 36", 42", 30", 36"

Refrigerators: 30" x 30", 30" x 36", 24" x 36", 24" x 42"

Sinks: 25", 33", 43"

Dishwasher

Washer

Kitchen Nook

Trash compactor

Dryer

Bathroom fixtures

Tubs and Showers: 32" x 60", 30" x 36", 36" x 48", 36" x 66", 36" x 36"

Sinks: 18" x 24", 14" x 18", 19", 20" x 24"

Toilet

Bidet

Vanities: 22" x 24", 22" x 60"

Windows (The number below each window is the Lindal part number)

	2'6"	2'6"	2'6"	2'6"	5'0"	5'0"	5'0"	5'0"	2'6"	5'0"
1'9"	850		853		846	850 853	853 853	851		
3'6"	849		855	856	847		849 856	855 856	864	867 (7'1")
5'4"	848	849/853	865	866	845	847/850/853	847/853/853	847/851		

Scale 3/32"=1'

▲Lindal Cedar Homes

Custom Home for:

Present address:
City/State/Province/Zip

Phone: Home/Business

Location of building site:

Instructions:

1. Lay the tracing paper on the plan of your choice or, if you are starting fresh, make rough sketches of the floorplan until you have the plan that's right for you.

2. Now, trace or draw free hand, first in pencil and then in ink when you are satisfied.

3. Note that the lines represent beam lines placed every 5' 4", which is our module width. Beams usually run in one direction only, but the grid has the lines in both directions so that you can lay out a variety of plans like a "T" and a "U", which have beams in two directions. Remember, in all cases the beams must run parallel to the ridge line.

4. Choose your windows from the window symbols on page 152 then trace and part number them on your plan accordingly.

5. If your home requires a staircase, make sure you have enough space for stairs and a landing. See Schedule 2.

6. Your bathrooms should be a minimum of 5' x 7' to accommodate the fixtures. Remember, it is also more economical to have a common plumbing wall for all fixtures. Trace the fixture symbols from page 152 on your plan. See pages 120-121 for more information.

7. Furniture can take up more room than expected, so use the furniture symbols from page 152 for space planning.

8. Check to make sure you have provided adequate storage area for all vestibules, laundry room, fuel storage, closets (bedroom, coat, linen) and exterior equipment storage.

9. Indicate north on the grid to enable our designers to offer site and passive solar assistance.

10. When you are satisfied with your home design, take this home planning grid, along with your site plan, to your local dealer for a free feasibility and cost analysis.

Schedules:

1. Typical roof details.

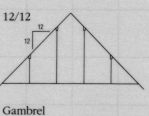

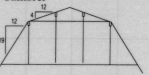

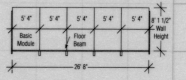

2. Stairs

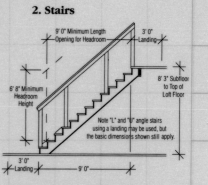

Note: Due to the printing process, some of the plans may not scale exactly to the dimensions. Use the plans for approximating sizes, and the dimensions for overall accuracy.

Instructions:

1. Lay the tracing paper on the plan of your choice or, if you are starting fresh, make rough sketches of the floorplan until you have the plan that's right for you.

2. Now, trace or draw free hand, first in pencil and then in ink when you are satisfied.

3. Note that the lines represent beam lines placed every 5' 4", which is our module width. Beams usually run in one direction only but the grid has the lines in both directions so that you can lay out a variety of plans like a "T" and a "L", which have beams in two directions. Remember, in all cases the beams must run parallel to the ridge line.

4. Choose your windows from the window symbols on page 152 then trace and part number them on your plan accordingly.

5. If your home requires a staircase, make sure you have enough space for stairs and a landing. See Schedule 2.

6. Your bathrooms should be a minimum of 5' x 7' to accommodate the fixtures. Remember it is also more economical to have a common plumbing wall for all fixtures. Trace the fixture symbols from page 152 on your plan. See pages 120-121 for more information.

7. Furniture can take up more room than expected, so use the furniture symbols from page 152 for space planning.

8. Check to make sure you have provided adequate storage area for all vestibules, laundry room, fuel storage, closets (bedroom, coat, linen) and exterior equipment storage.

9. Indicate north on the grid to enable our designers to offer site and passive solar assistance.

10. When you are satisfied with your home design, take this home planning grid, along with your site plan, to your local dealer for a free feasibility and cost analysis.

Schedules:

1. Typical roof details.

4/12

8/12

12/12

Gambrel

2. Stairs

Scale 3/32" = 1'
5' 4" = Lindal Module

DESIGNER SERIES

SUNDOWNER

- 3 Bedrooms
- 2 Bathrooms
- Master bedroom on second floor
- 2016 sq. ft.
- First Floor: 1425 sq. ft.
- Other Floor: 591 sq. ft.
- Overall Size: 51' x 33'

Photos of Designers are shown on pages 34-39.

Shading indicates skylight or sunroom.

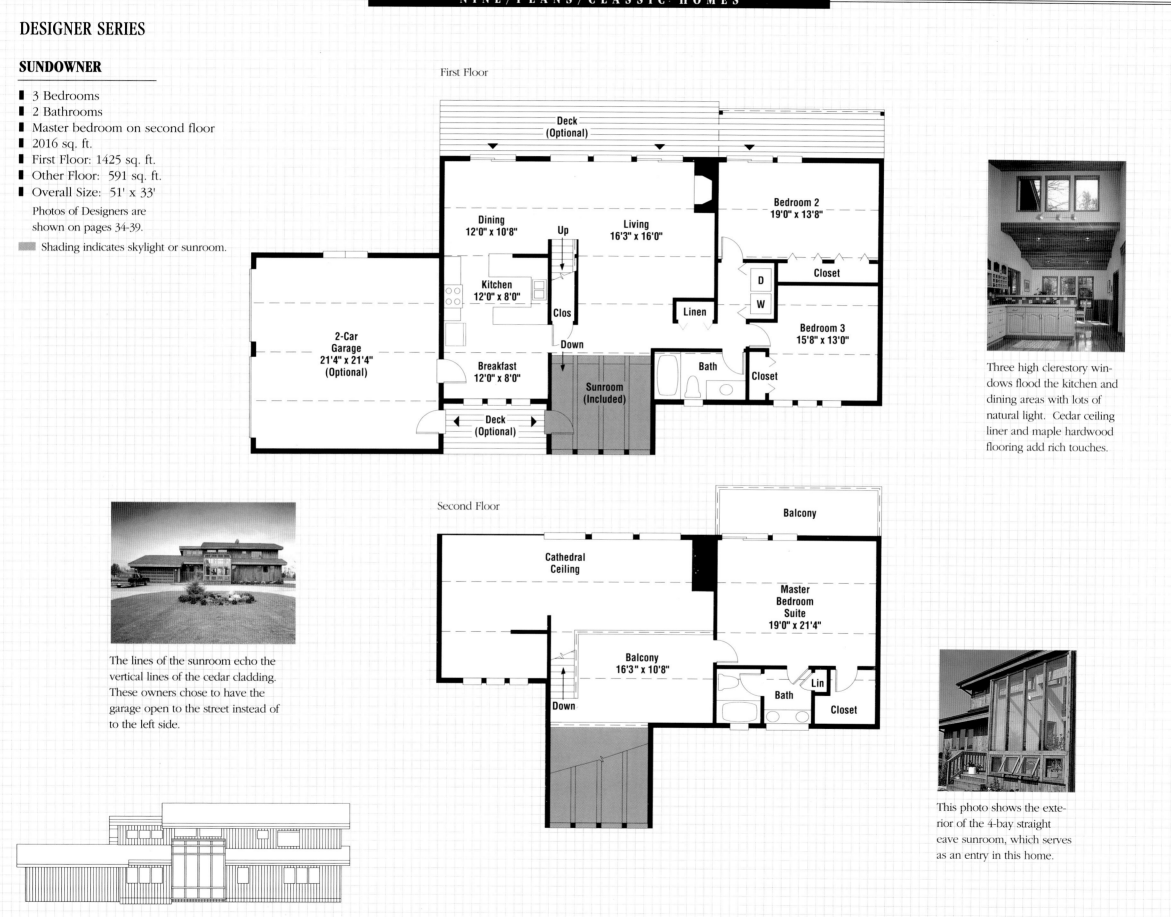

First Floor

Deck (Optional)

Dining 12'0" x 10'8"

Up

Living 16'3" x 16'0"

Bedroom 2 19'0" x 13'8"

Closet

Kitchen 12'0" x 8'0"

Clos

D

W

Linen

Bedroom 3 15'8" x 13'0"

Down

2-Car Garage 21'4" x 21'4" (Optional)

Breakfast 12'0" x 8'0"

Sunroom (Included)

Bath

Closet

Deck (Optional)

Three high clerestory windows flood the kitchen and dining areas with lots of natural light. Cedar ceiling liner and maple hardwood flooring add rich touches.

Second Floor

Balcony

Cathedral Ceiling

Master Bedroom Suite 19'0" x 21'4"

Balcony 16'3" x 10'8"

Lin

Bath

Down

Closet

The lines of the sunroom echo the vertical lines of the cedar cladding. These owners chose to have the garage open to the street instead of to the left side.

This photo shows the exterior of the 4-bay straight eave sunroom, which serves as an entry in this home.

CITY SERIES

BOSTON

- 3 Bedrooms
- 2 1/2 Bathrooms
- Master bedroom on second floor
- 1624 sq. ft.
- First Floor: 910 sq. ft.
- Other Floor: 714 sq. ft.
- Overall Size: 35' x 31'

Photos of Cities are
shown on pages 40-43.

Shading indicates skylight or sunroom.

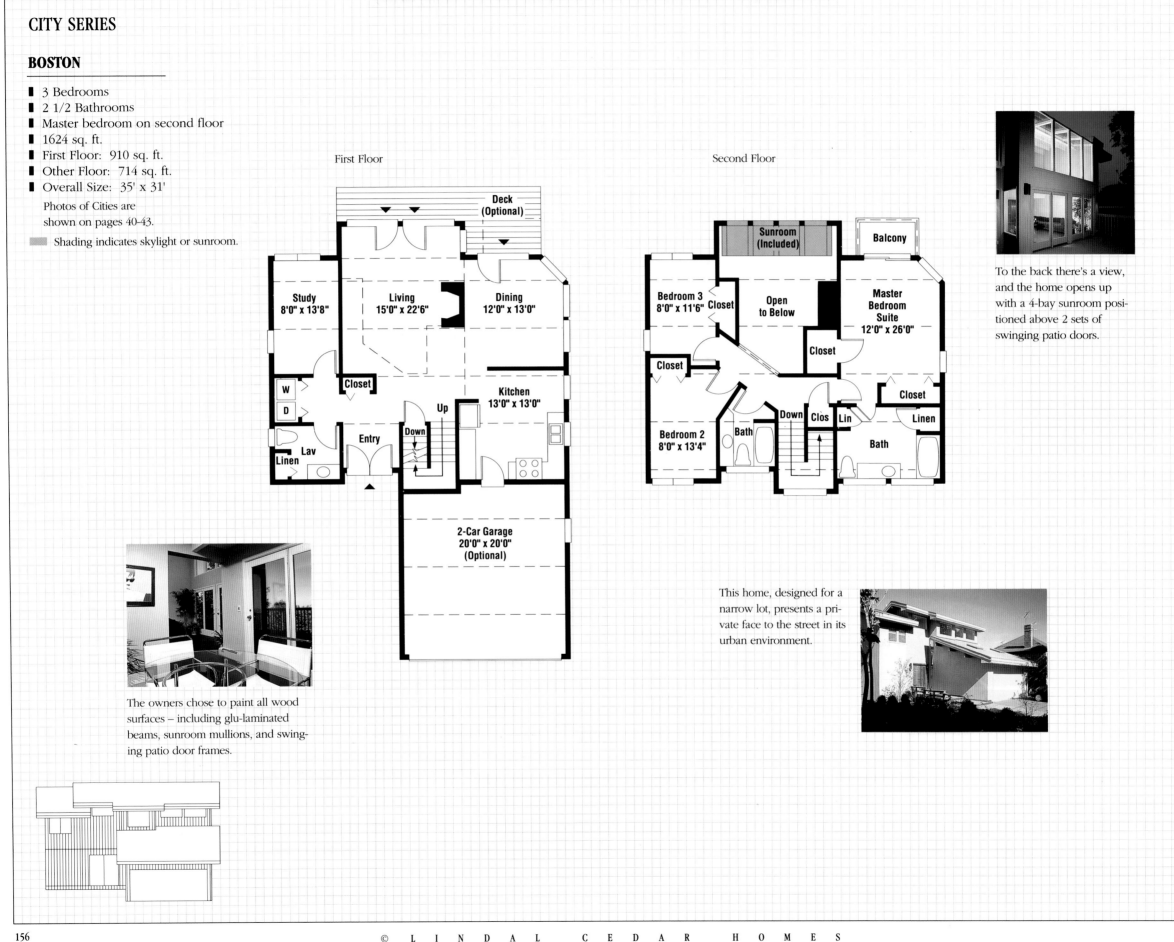

First Floor

**Deck
(Optional)**

Second Floor

**Sunroom
(Included)**

Balcony

Study
8'0" x 13'8"

Living
15'0" x 22'6"

Dining
12'0" x 13'0"

Bedroom 3
8'0" x 11'6"

Closet

**Open
to Below**

**Master
Bedroom
Suite**
12'0" x 26'0"

Closet

Closet

Closet

W

D

Closet

Kitchen
13'0" x 13'0"

Closet

Closet

Linen

Bedroom 2
8'0" x 13'4"

Bath

Down

Clos

Lin

Linen

Up

Bath

Lav

Linen

Entry

Down

2-Car Garage
20'0" x 20'0"
(Optional)

To the back there's a view,
and the home opens up
with a 4-bay sunroom positioned above 2 sets of
swinging patio doors.

The owners chose to paint all wood
surfaces – including glu-laminated
beams, sunroom mullions, and swinging patio door frames.

This home, designed for a
narrow lot, presents a private face to the street in its
urban environment.

CITY SERIES

TORONTO

- 4 Bedrooms
- 3 Bathrooms
- Master bedroom on second floor
- 2384 sq. ft.
- First Floor: 1330 sq. ft.
- Other Floor: 1054 sq. ft.
- Overall Size: 43' x 37'

Photos of Cities are shown on pages 40-43.

Shading indicates skylight or sunroom.

Positioning the sunroom over opening doors of a matching width, makes a dramatic statement. Note: The owners substituted swinging patio doors for sliding glass doors and added a window.

Upstairs in Bedroom 4, the owners also substituted swinging patio doors for standard sliding glass doors and added another window.

This home has a walk-out daylight basement. From the back corner of the site, you can see the three stories and two levels of balconies.

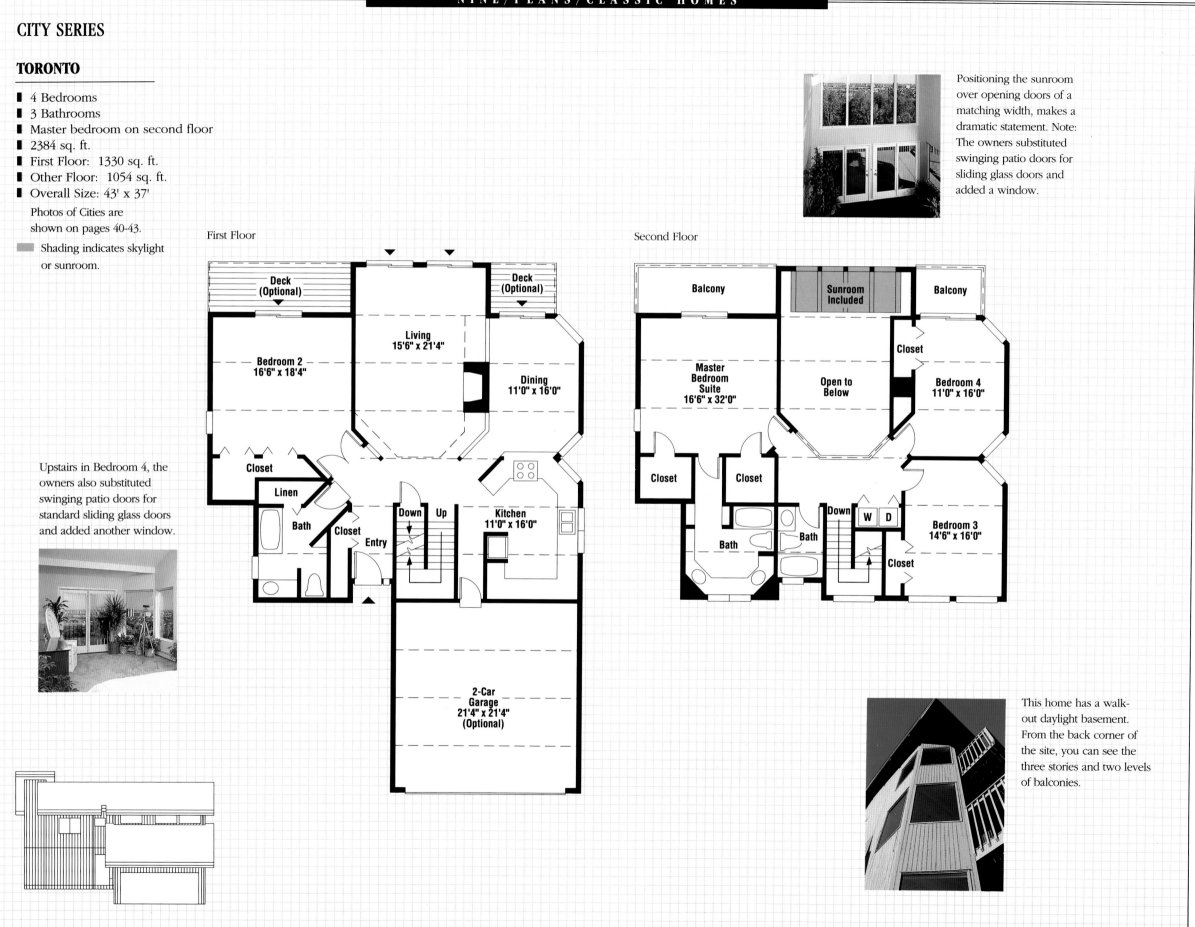

First Floor

Deck (Optional)

Deck (Optional)

Living 15'6" x 21'4"

Bedroom 2 16'6" x 18'4"

Dining 11'0" x 16'0"

Closet

Linen

Bath

Closet

Entry

Down Up

Kitchen 11'0" x 16'0"

2-Car Garage 21'4" x 21'4" (Optional)

Second Floor

Balcony

Sunroom Included

Balcony

Closet

Master Bedroom Suite 16'6" x 32'0"

Open to Below

Bedroom 4 11'0" x 16'0"

Closet Closet

Down W D

Bath Bath

Bedroom 3 14'6" x 16'0"

Closet

CITY SERIES

PHILADELPHIA

- 4 Bedrooms
- 3 Bathrooms
- Master bedroom on second floor
- 2421 sq. ft.
- First Floor: 1439 sq. ft.
- Other Floor: 982 sq. ft.
- Overall Size: 43' x 38'

Photos of Cities are shown on pages 40-43.

▨ Shading indicates skylight or sunroom.

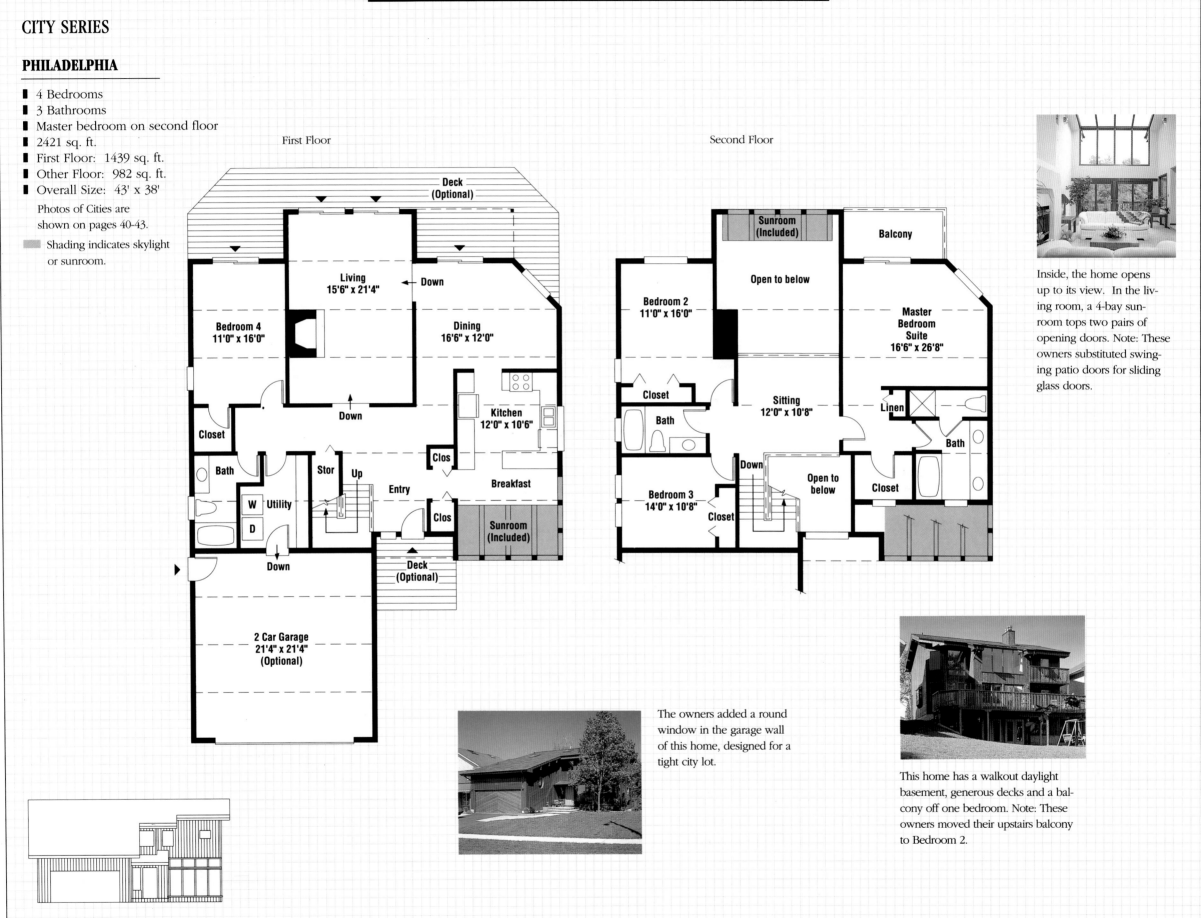

First Floor

Deck (Optional)

Living 15'6" x 21'4"

← Down

Bedroom 4 11'0" x 16'0"

Dining 16'6" x 12'0"

Down

Kitchen 12'0" x 10'6"

Closet

Bath

Stor

Up

Clos

Breakfast

W D Utility

Entry

Clos

Sunroom (Included)

Down

Deck (Optional)

2 Car Garage 21'4" x 21'4" (Optional)

Second Floor

Sunroom (Included)

Balcony

Open to below

Bedroom 2 11'0" x 16'0"

Master Bedroom Suite 16'6" x 26'8"

Closet

Bath

Sitting 12'0" x 10'8"

Linen

Down

Bath

Bedroom 3 14'0" x 10'8"

Closet

Open to below

Closet

Inside, the home opens up to its view. In the living room, a 4-bay sunroom tops two pairs of opening doors. Note: These owners substituted swinging patio doors for sliding glass doors.

The owners added a round window in the garage wall of this home, designed for a tight city lot.

This home has a walkout daylight basement, generous decks and a balcony off one bedroom. Note: These owners moved their upstairs balcony to Bedroom 2.

EXECUTIVE SERIES

BELAIR

- 2 Bedrooms
- 2 Bathrooms
- Master bedroom on second floor
- 1760 sq. ft.
- First Floor: 1220 sq. ft.
- Other Floor: 540 sq. ft.
- Overall Size: 47' x 32'

Photos of Executives are shown on pages 44-47.

Stepped-down roofs add visual interest to this suburban home. Note: The owners preferred a double door entry, and extended the garage wing.

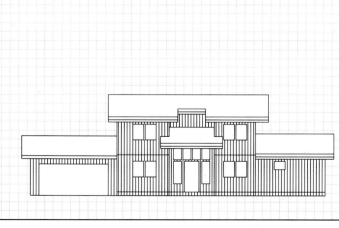

First Floor

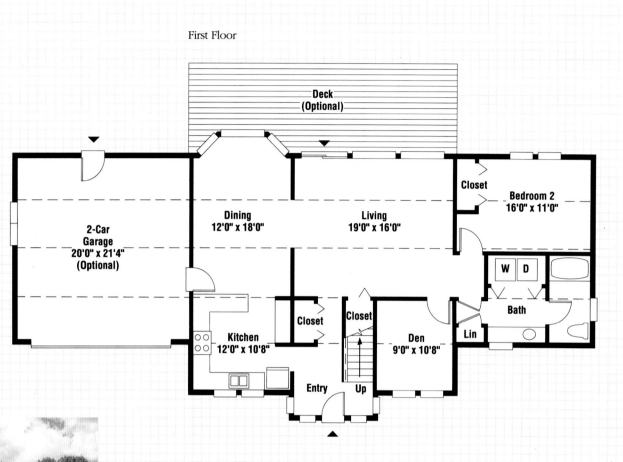

Deck (Optional)

Closet

Bedroom 2
16'0" x 11'0"

Dining
12'0" x 18'0"

Living
19'0" x 16'0"

2-Car Garage
20'0" x 21'4"
(Optional)

W   D

Bath

Closet   Closet

Lin

Den
9'0" x 10'8"

Kitchen
12'0" x 10'8"

Entry   Up

Second Floor

Open to Below

Master Bedroom Suite
12'0" x 26'8"

Balcony

Closet

Dn

Lin

Study
9'0" x 16'0"

Bath

Open to Below

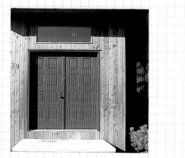

The standard front door in this plan is a steel insulated entry door system with weatherstripping (R-14). Optional double doors make a welcoming entry.

EXECUTIVE SERIES

GEORGETOWN

- 3 Bedrooms
- 2 1/2 Bathrooms
- Master bedroom on first floor
- 2223 sq. ft.
- First Floor: 1627 sq. ft.
- Other Floor: 596 sq. ft.
- Overall Size: 64' x 32'

Photos of Executives are shown on pages 44-47.

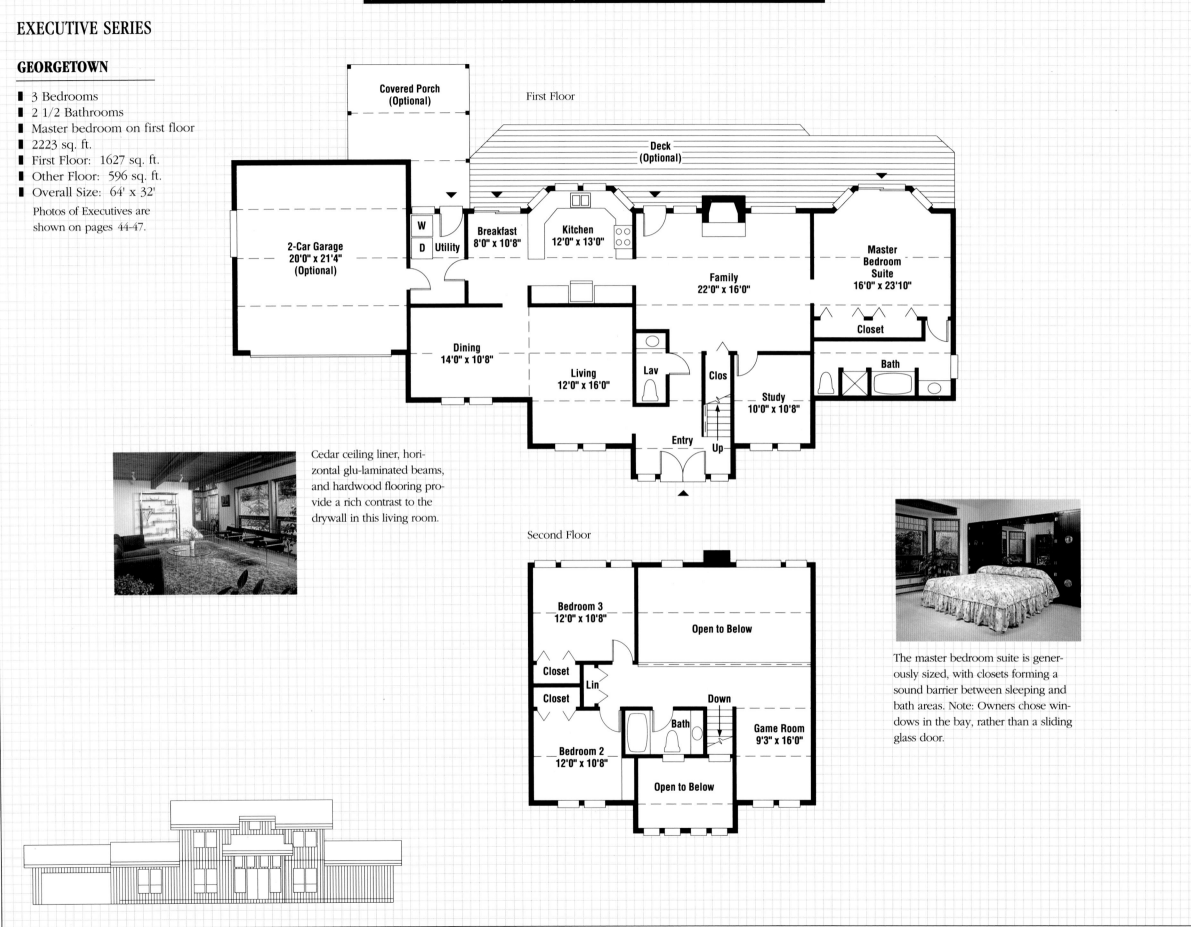

First Floor

Covered Porch
(Optional)

Deck
(Optional)

W
D Utility

Breakfast
8'0" x 10'8"

Kitchen
12'0" x 13'0"

Master
Bedroom
Suite
16'0" x 23'10"

2-Car Garage
20'0" x 21'4"
(Optional)

Family
22'0" x 16'0"

Closet

Bath

Dining
14'0" x 10'8"

Living
12'0" x 16'0"

Lav

Clos

Study
10'0" x 10'8"

Entry

Up

Cedar ceiling liner, horizontal glu-laminated beams, and hardwood flooring provide a rich contrast to the drywall in this living room.

Second Floor

Bedroom 3
12'0" x 10'8"

Open to Below

Closet

Closet

Lin

Down

Bath

Game Room
9'3" x 16'0"

Bedroom 2
12'0" x 10'8"

Open to Below

The master bedroom suite is generously sized, with closets forming a sound barrier between sleeping and bath areas. Note: Owners chose windows in the bay, rather than a sliding glass door.

EXECUTIVE SERIES

SCARSDALE

- 4 Bedrooms
- 3 1/2 Bathrooms
- Master bedroom on first floor
- 3394 sq. ft.
- First Floor: 2466 sq. ft.
- Other Floor: 928 sq. ft.
- Overall Size: 68' x 43'

Photos of Executives are
shown on pages 44-47.

The master bedroom and
garage are in the 1-story
wings of this spacious
home. Note: Glass astrals
can be substituted for wood
in the gable ends.

In the master bedroom, a
sliding glass door opens
to the deck. Glass astrals
add light.

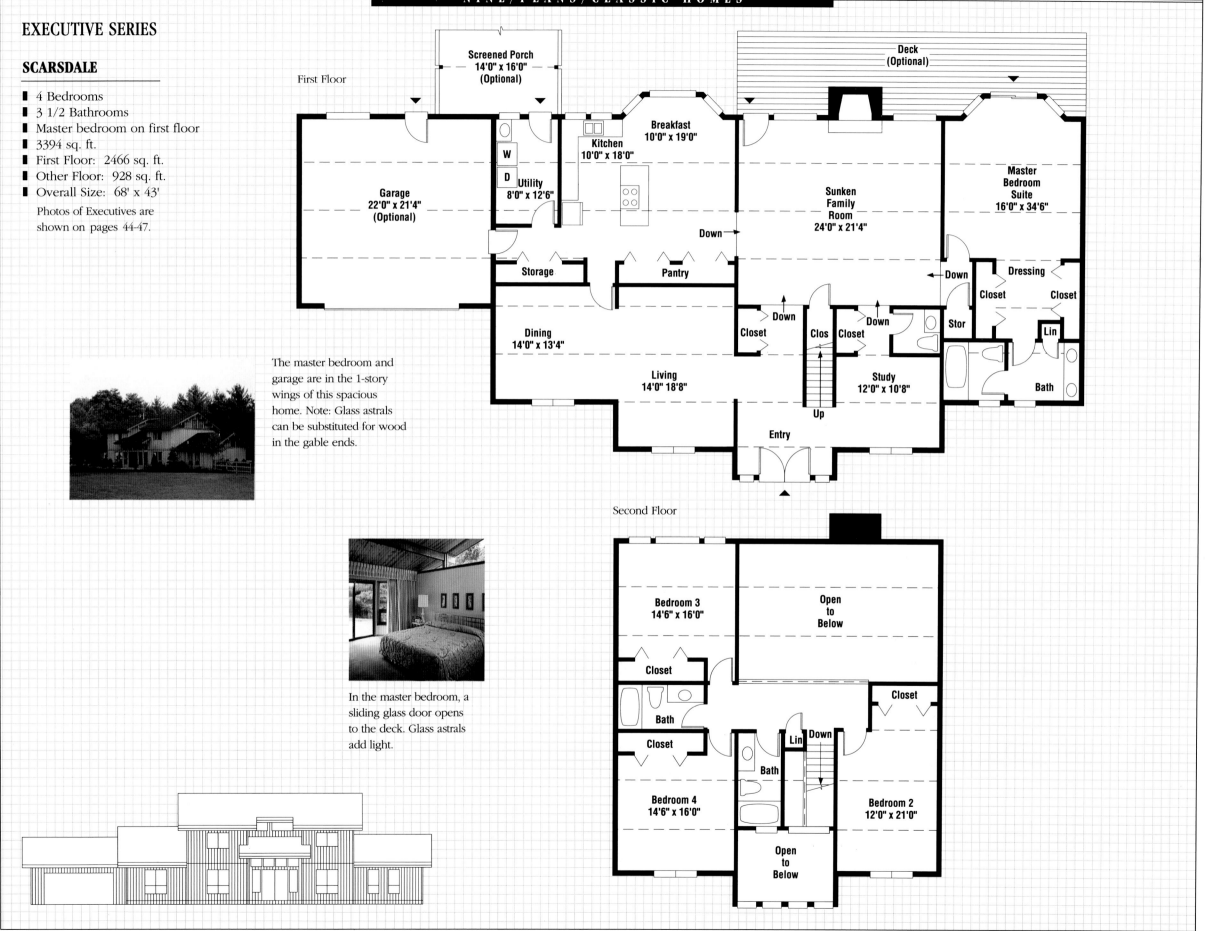

First Floor

Screened Porch
14'0" x 16'0"
(Optional)

Deck
(Optional)

Breakfast
10'0" x 19'0"

Kitchen
10'0" x 18'0"

W

D

Utility
8'0" x 12'6"

Garage
22'0" x 21'4"
(Optional)

Sunken
Family
Room
24'0" x 21'4"

Master
Bedroom
Suite
16'0" x 34'6"

Down

Storage

Pantry

Down

Down

Dressing

Closet

Closet

Down

Stor

Lin

Dining
14'0" x 13'4"

Closet

Clos

Closet

Bath

Living
14'0" 18'8"

Study
12'0" x 10'8"

Up

Entry

Second Floor

Bedroom 3
14'6" x 16'0"

Open
to
Below

Closet

Closet

Bath

Closet

Lin

Down

Closet

Bath

Bedroom 4
14'6" x 16'0"

Bedroom 2
12'0" x 21'0"

Open
to
Below

CONTINENTAL SERIES

TUSCANY

- 3 Bedrooms
- 2 Bathrooms
- Master bedroom on second floor
- 1615 sq. ft.
- First floor: 1035 sq. ft.
- Other floor: 580 sq. ft.
- Overall Size: 37' x 31'

Photos of Continentals are shown on pages 48-49.

■ Shading indicates skylight or sunroom.

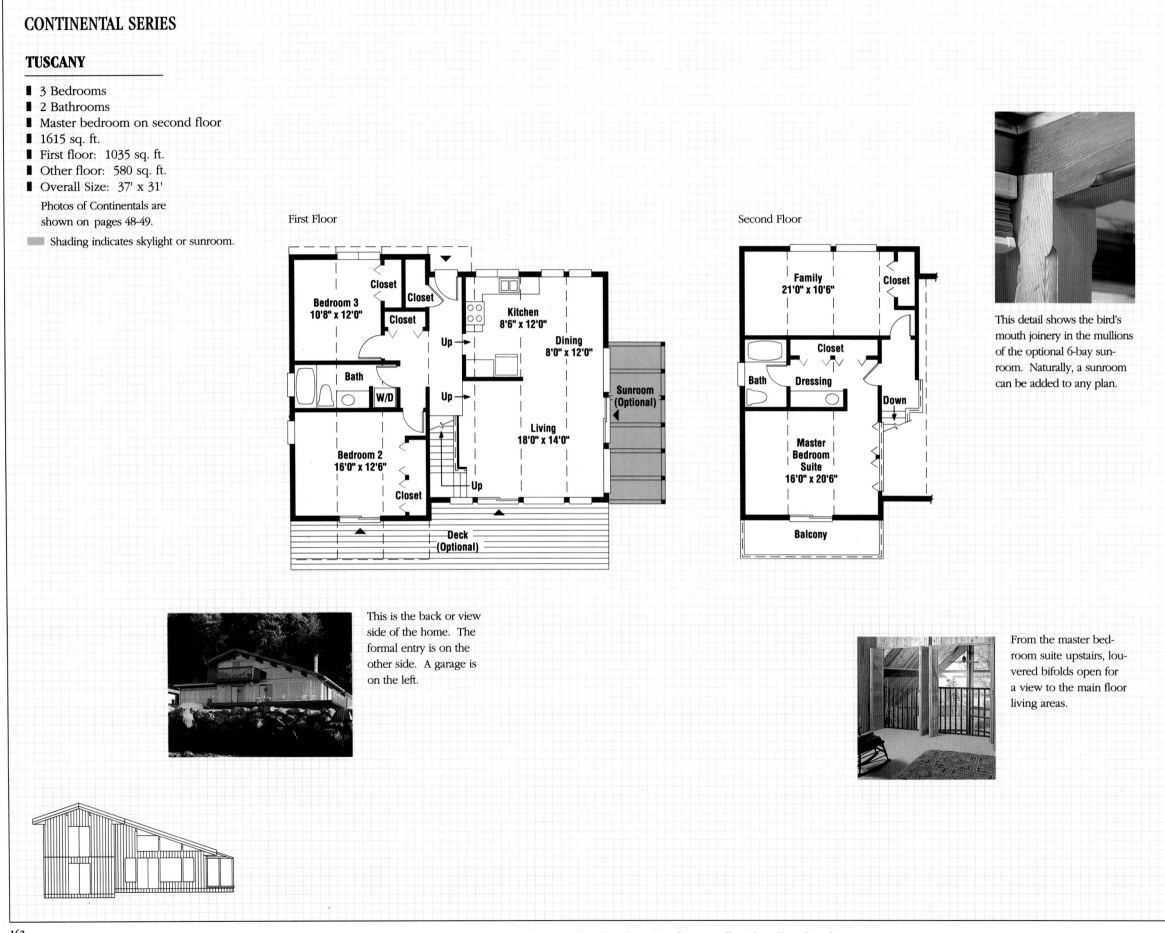

First Floor

Closet
Bedroom 3
10'8" x 12'0"
Closet
Closet
Bath
W/D
Bedroom 2
16'0" x 12'6"
Closet
Up
Up
Up
Kitchen
8'6" x 12'0"
Dining
8'0" x 12'0"
Living
18'0" x 14'0"
Sunroom
(Optional)
Deck
(Optional)

Second Floor

Family
21'0" x 10'6"
Closet
Closet
Bath
Dressing
Down
Master
Bedroom
Suite
16'0" x 20'6"
Balcony

This detail shows the bird's mouth joinery in the mullions of the optional 6-bay sunroom. Naturally, a sunroom can be added to any plan.

This is the back or view side of the home. The formal entry is on the other side. A garage is on the left.

From the master bedroom suite upstairs, louvered bifolds open for a view to the main floor living areas.

CONTINENTAL SERIES

BAVARIA

- 2+ Bedrooms
- 2 3/4 Bathrooms
- Master bedroom on second floor
- 2166 sq. ft.
- First floor: 1399 sq. ft.
- Other floor: 767 sq. ft.
- Overall Size: 43' x 44'

Photos of Continentals are shown on pages 48-49.

Shading indicates skylight or sunroom.

In this plan, the living room projects on the right and the breakfast room on the left.

This is the back or view side of the home. The formal entry and garage are on the other side.

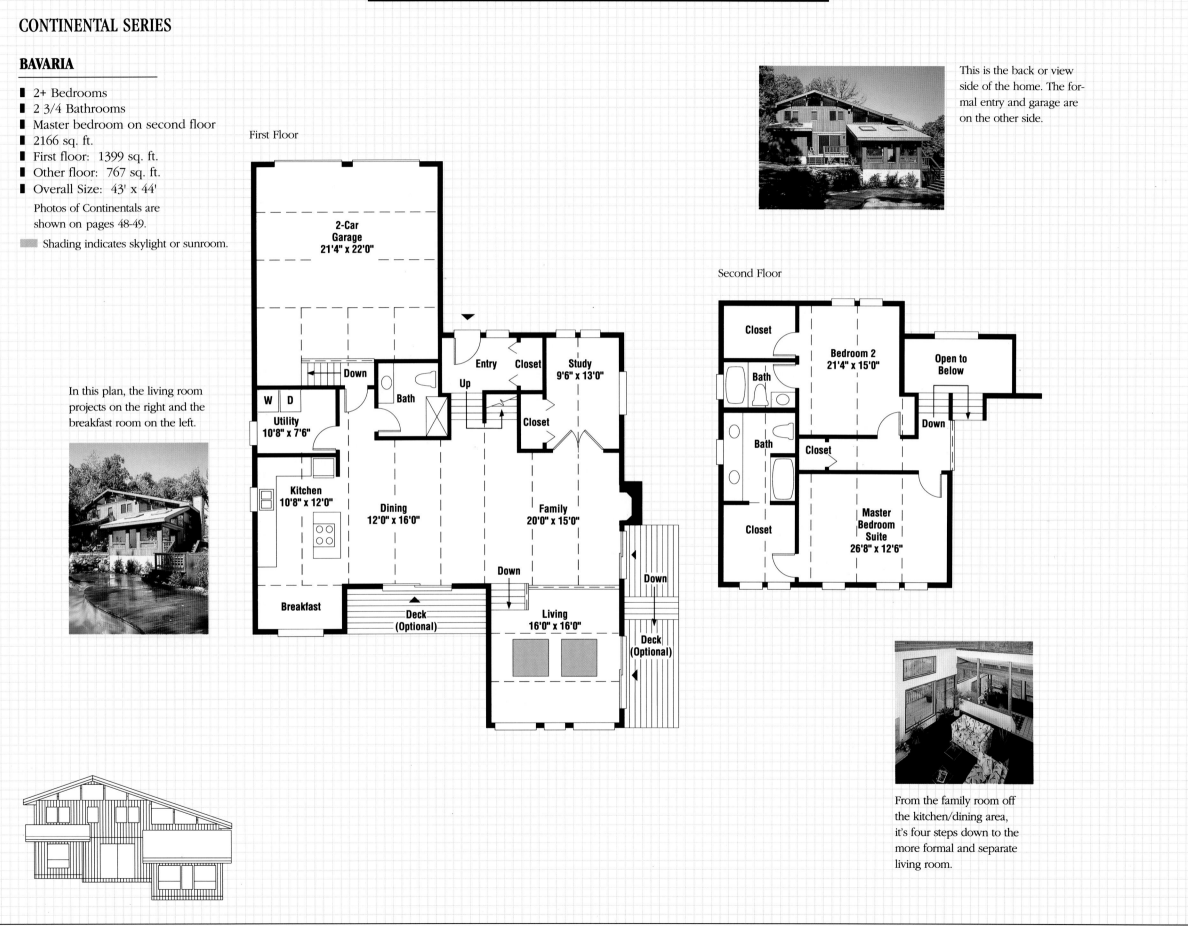

First Floor

2-Car Garage 21'4" x 22'0"

Down

W D

Utility 10'8" x 7'6"

Kitchen 10'8" x 12'0"

Bath

Entry   Closet   **Study** 9'6" x 13'0"

Up

Closet

Dining 12'0" x 16'0"

Family 20'0" x 15'0"

Breakfast

Down

Deck (Optional)

Down

Living 16'0" x 16'0"

Down

Deck (Optional)

Second Floor

Closet

Bath

Bedroom 2 21'4" x 15'0"

Open to Below

Bath

Closet

Down

Closet

Master Bedroom Suite 26'8" x 12'6"

From the family room off the kitchen/dining area, it's four steps down to the more formal and separate living room.

CONTINENTAL SERIES

BURGUNDY

- 4 Bedrooms
- 3 Bathrooms
- Master bedroom on second floor
- 2273 sq. ft.
- First floor: 1450 sq. ft.
- Other floor: 823 sq. ft.
- Overall Size: 43' x 40'

Photos of Continentals are shown on pages 48-49.

This is the front or street side of the home. Note how the upper floor projects over the lower floor on the left.

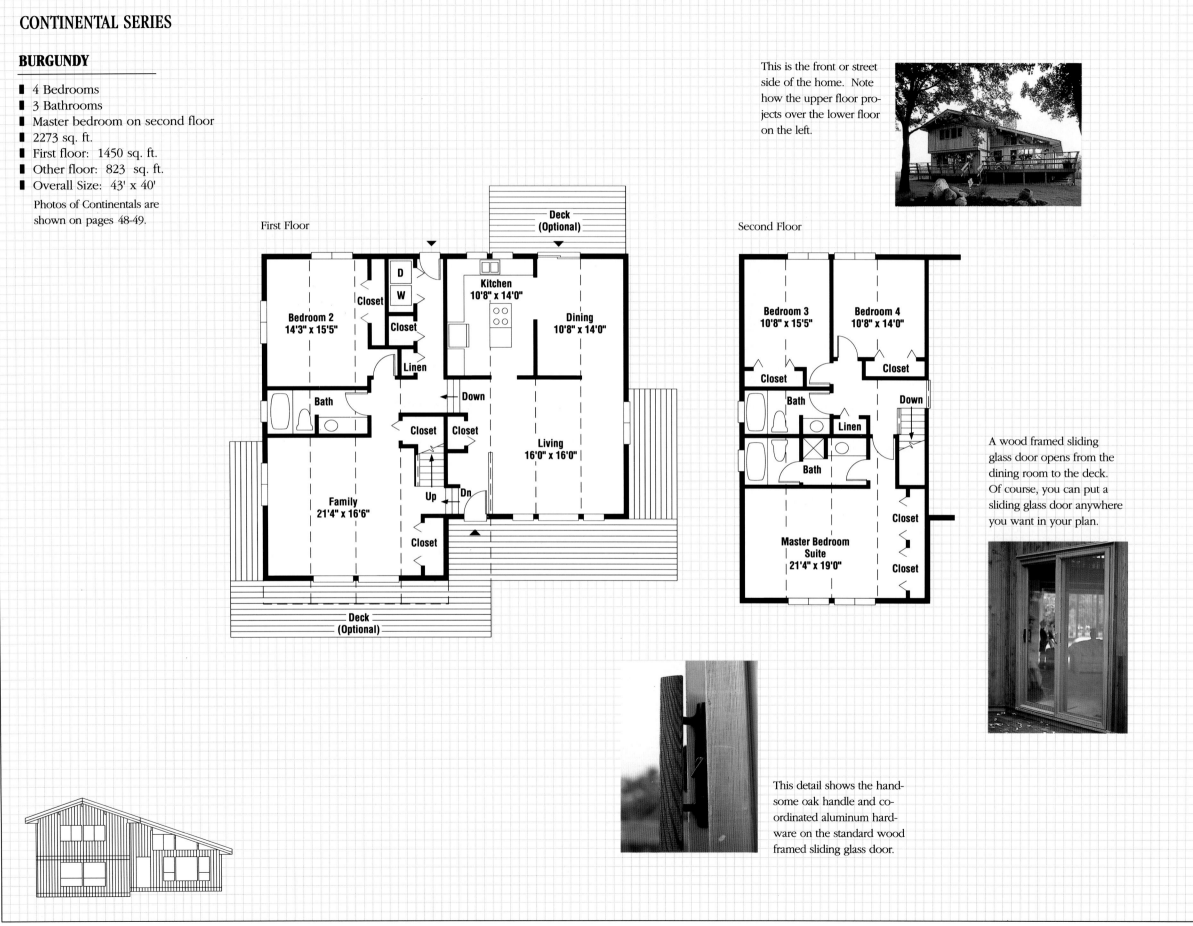

First Floor

Deck (Optional)

Kitchen 10'8" x 14'0"

Dining 10'8" x 14'0"

D / W

Closet

Closet

Bedroom 2 14'3" x 15'5"

Linen

Bath

Down

Living 16'0" x 16'0"

Closet

Closet

Up

Dn

Family 21'4" x 16'6"

Closet

Deck (Optional)

Second Floor

Bedroom 3 10'8" x 15'5"

Bedroom 4 10'8" x 14'0"

Closet

Closet

Bath

Linen

Down

Bath

Closet

Master Bedroom Suite 21'4" x 19'0"

Closet

A wood framed sliding glass door opens from the dining room to the deck. Of course, you can put a sliding glass door anywhere you want in your plan.

This detail shows the handsome oak handle and coordinated aluminum hardware on the standard wood framed sliding glass door.

INTERNATIONAL SERIES

MADRID

- 3 Bedrooms
- 2 Bathrooms
- 1467 sq. ft.
- Overall Size: 43' x 49'

Photos of Internationals are shown on pages 50-51.

A screened porch is ideal for hot climates where bugs can be a nuisance. The Madrid features an optional porch with cedar ceiling liner and decking.

Our standard interior doors are solid core Red Oak with oil rubbed bronze finish hardware (U.S.) and polished brass (Canada).

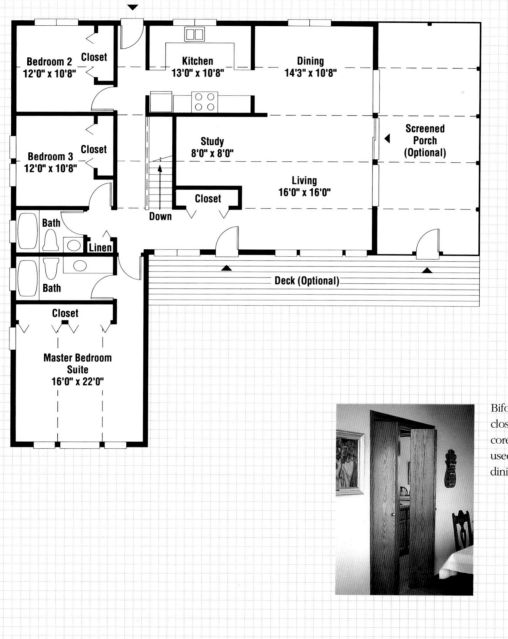

Bedroom 2
12'0" x 10'8" Closet

Kitchen
13'0" x 10'8"

Dining
14'3" x 10'8"

Bedroom 3
12'0" x 10'8" Closet

Study
8'0" x 8'0"

Screened Porch (Optional)

Closet

Living
16'0" x 16'0"

Down

Bath

Linen

Bath

Closet

Master Bedroom Suite
16'0" x 22'0"

Deck (Optional)

Bifolds are often used in closets. Here, flush hollow core Red oak bifolds are used between a kitchen and dining room.

INTERNATIONAL SERIES

TOKYO

- 3 Bedrooms
- 2 Bathrooms
- 2627 sq. ft.
- Overall Size: 66' x 56'

Photos of Internationals are shown on pages 50-51.

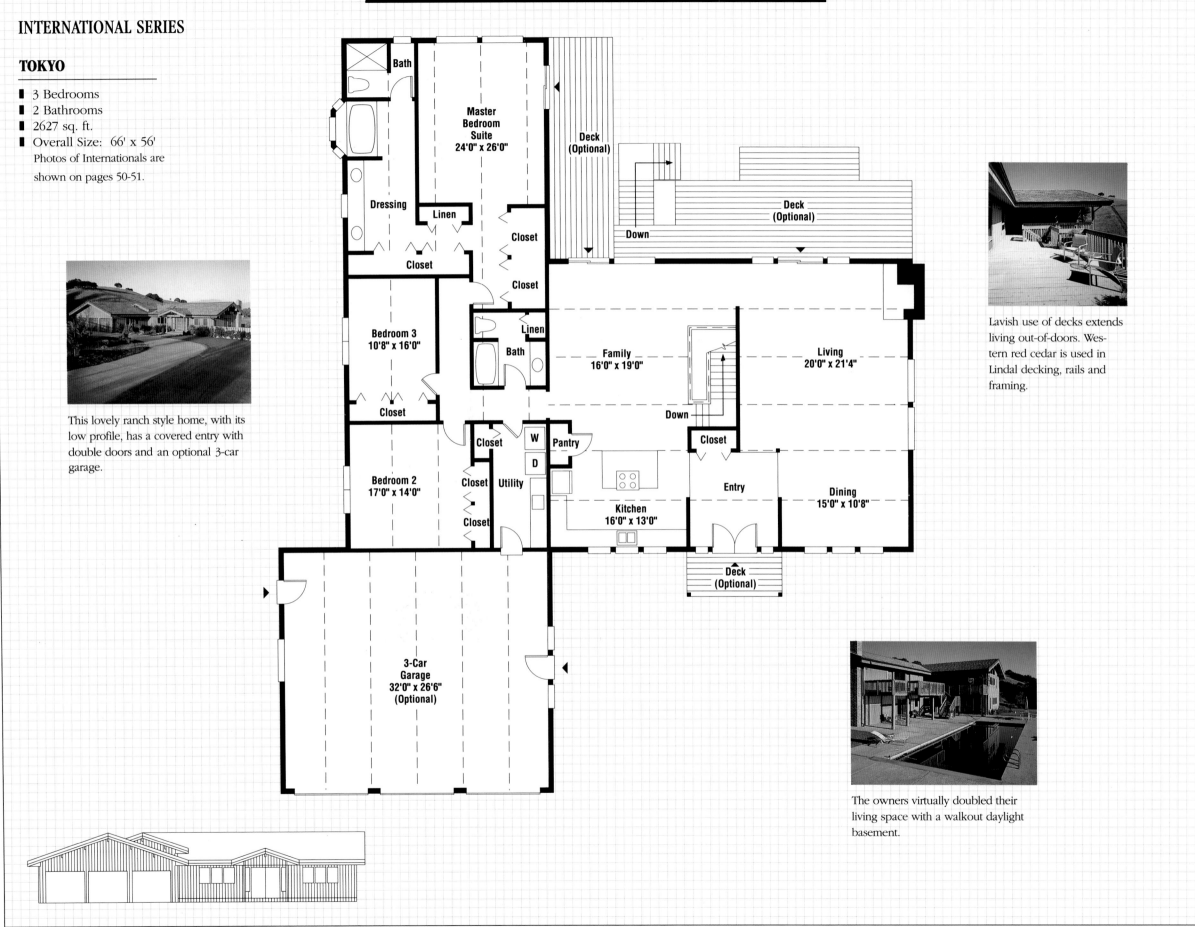

Bath

Master Bedroom Suite
24'0" x 26'0"

Deck (Optional)

Deck (Optional)

Down

Dressing

Linen

Closet

Closet

Closet

Linen

Bedroom 3
10'8" x 16'0"

Closet

Bath

Family
16'0" x 19'0"

Down

Living
20'0" x 21'4"

Closet

Closet

W
D

Pantry

Closet

Entry

Dining
15'0" x 10'8"

Bedroom 2
17'0" x 14'0"

Closet

Utility

Kitchen
16'0" x 13'0"

Closet

Deck (Optional)

3-Car Garage
32'0" x 26'6"
(Optional)

This lovely ranch style home, with its low profile, has a covered entry with double doors and an optional 3-car garage.

Lavish use of decks extends living out-of-doors. Western red cedar is used in Lindal decking, rails and framing.

The owners virtually doubled their living space with a walkout daylight basement.

SIGNATURE SERIES

LANDMARK

- 3 Bedrooms
- 2 Bathrooms
- Master bedroom on second floor
- 1876 sq. ft.
- First Floor: 1216 sq. ft.
- Other Floor: 660 sq. ft.
- Overall Size: 43' x 32'

Photos of Signatures are shown on pages 52-53.

Shading indicates skylight or sunroom.

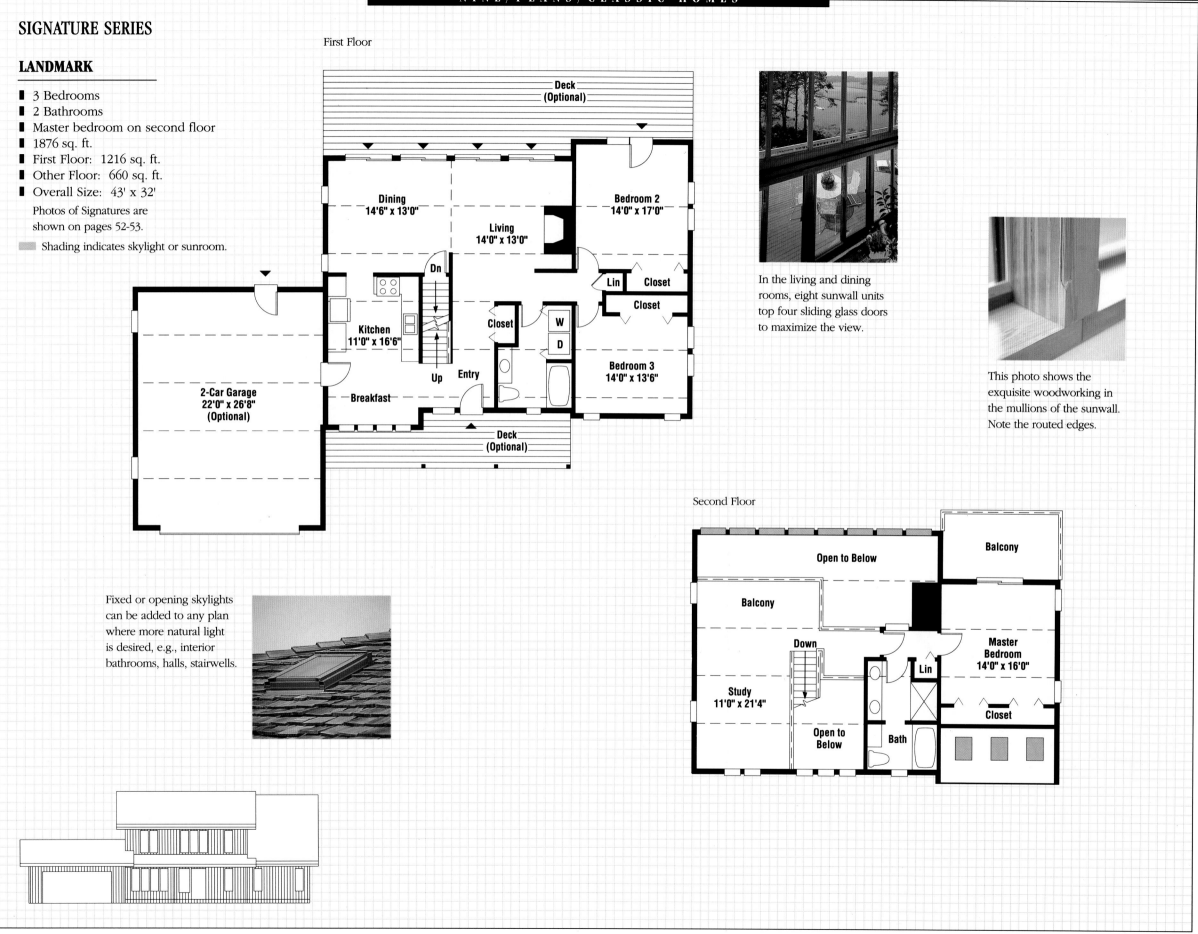

First Floor

Deck (Optional)

Dining 14'6" x 13'0"

Living 14'0" x 13'0"

Bedroom 2 14'0" x 17'0"

Dn

Lin Closet

Closet

Kitchen 11'0" x 16'6"

Closet

W D

Bedroom 3 14'0" x 13'6"

Up Entry

Breakfast

2-Car Garage 22'0" x 26'8" (Optional)

Deck (Optional)

In the living and dining rooms, eight sunwall units top four sliding glass doors to maximize the view.

This photo shows the exquisite woodworking in the mullions of the sunwall. Note the routed edges.

Fixed or opening skylights can be added to any plan where more natural light is desired, e.g., interior bathrooms, halls, stairwells.

Second Floor

Open to Below

Balcony

Balcony

Down

Lin

Master Bedroom 14'0" x 16'0"

Study 11'0" x 21'4"

Open to Below

Bath

Closet

SIGNATURE SERIES

HALLMARK

- 3 Bedrooms
- 2 1/2 Bathrooms
- Master bedroom on first floor
- 2325 sq. ft.
- First Floor: 1513 sq. ft.
- Other Floor: 812 sq. ft.
- Overall Size: 50' x 35'

Photos of Signatures are shown on pages 52-53.

Shading indicates skylight or sunroom.

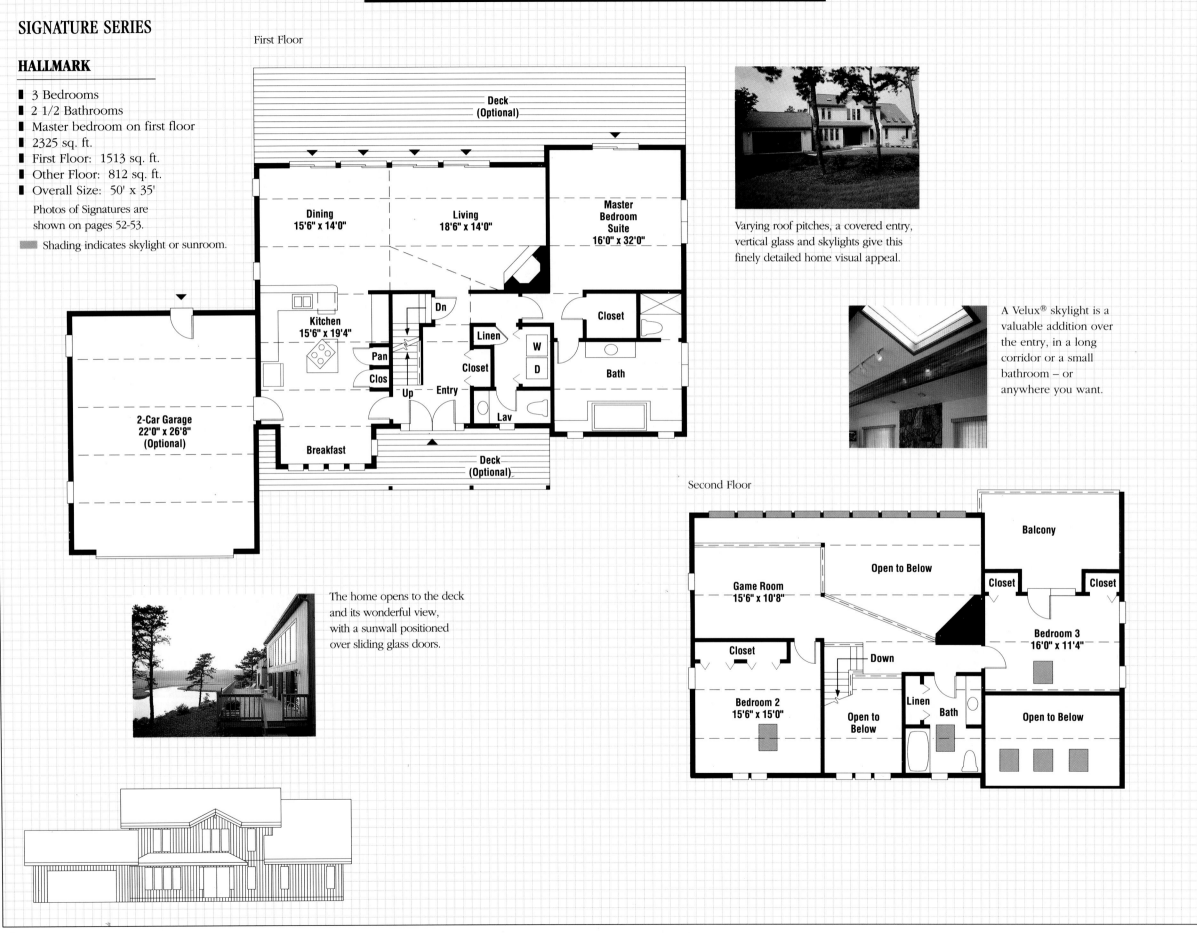

First Floor

Dining
15'6" x 14'0"

Living
18'6" x 14'0"

Master
Bedroom
Suite
16'0" x 32'0"

Deck
(Optional)

Kitchen
15'6" x 19'4"

Closet

Dn

Linen

W
D

Pan

Closet

Bath

Clos

Up   Entry

Lav

Breakfast

Deck
(Optional)

2-Car Garage
22'0" x 26'8"
(Optional)

Varying roof pitches, a covered entry, vertical glass and skylights give this finely detailed home visual appeal.

A Velux® skylight is a valuable addition over the entry, in a long corridor or a small bathroom – or anywhere you want.

The home opens to the deck and its wonderful view, with a sunwall positioned over sliding glass doors.

Second Floor

Balcony

Game Room
15'6" x 10'8"

Open to Below

Closet

Closet

Closet

Bedroom 3
16'0" x 11'4"

Closet

Down

Bedroom 2
15'6" x 15'0"

Open to Below

Linen

Bath

Open to Below

Open to Below

EMBASSY SERIES

CONSUL

- 3 Bedrooms
- 2 1/2 Bathrooms
- Master bedroom on second floor
- 2954 sq. ft.
- First Floor: 1613 sq. ft.
- Other Floor: 1341 sq. ft.
- Overall Size: 41' x 44'

Photos of Embassies are shown on pages 54-55.

▨ Shading indicates skylight or sunroom.

AMBASSADOR

- 4 Bedrooms
- 3 1/2 Bathrooms
- Master bedroom on first floor
- 3495 sq. ft.
- First Floor: 2154 sq. ft.
- Other Floor: 1341 sq. ft.
- Overall Size: 56' x 44'

First Floor

Deck (Optional)

Kitchen 14'6" x 13'0"

Family 27'0" x 16'0"

Bath

Pantry

Closet

Closet

Cl

Closet

Dining 14'6" x 13'8"

Living 14'6" x 21'4'

D   W

Up

Sunroom (Included)

Entry

Second Floor

Bedroom 2 14'6" x 16'0"

Open to Below

Master Bedroom Suite 14'6" x 37'4"

Closet

Bath

Closet

Game Room 12'3" x 23'8"

Down

Closet

Closet

Closet

Closet

Bath

Bedroom 3 14'6" x 16'0"

Bath

Deck (Optional)

Master Bedroom Suite 14'6" x 37'4"

Closet

Closet

Closet

Closet

Lav

Bath

This wing can be added to the Consul now, or later, to provide more living area and put the master bedroom suite on the main floor in its own private wing.

EMBASSY SERIES

DIPLOMAT

- 4 Bedrooms
- 3 1/2 Bathrooms
- Master bedroom on first floor
- 4420 sq. ft.
- First Floor: 3148 sq. ft.

- Other Floor: 1272 sq. ft.
- Overall Size: 72' x 51'

Photos of Embassies are
shown on pages 54-55.

▨ Shading indicates skylight or sunroom.

Most Lindal homes come
with wide overhangs. Here,
the absence of eaves is reminiscent of saltbox design.

First Floor

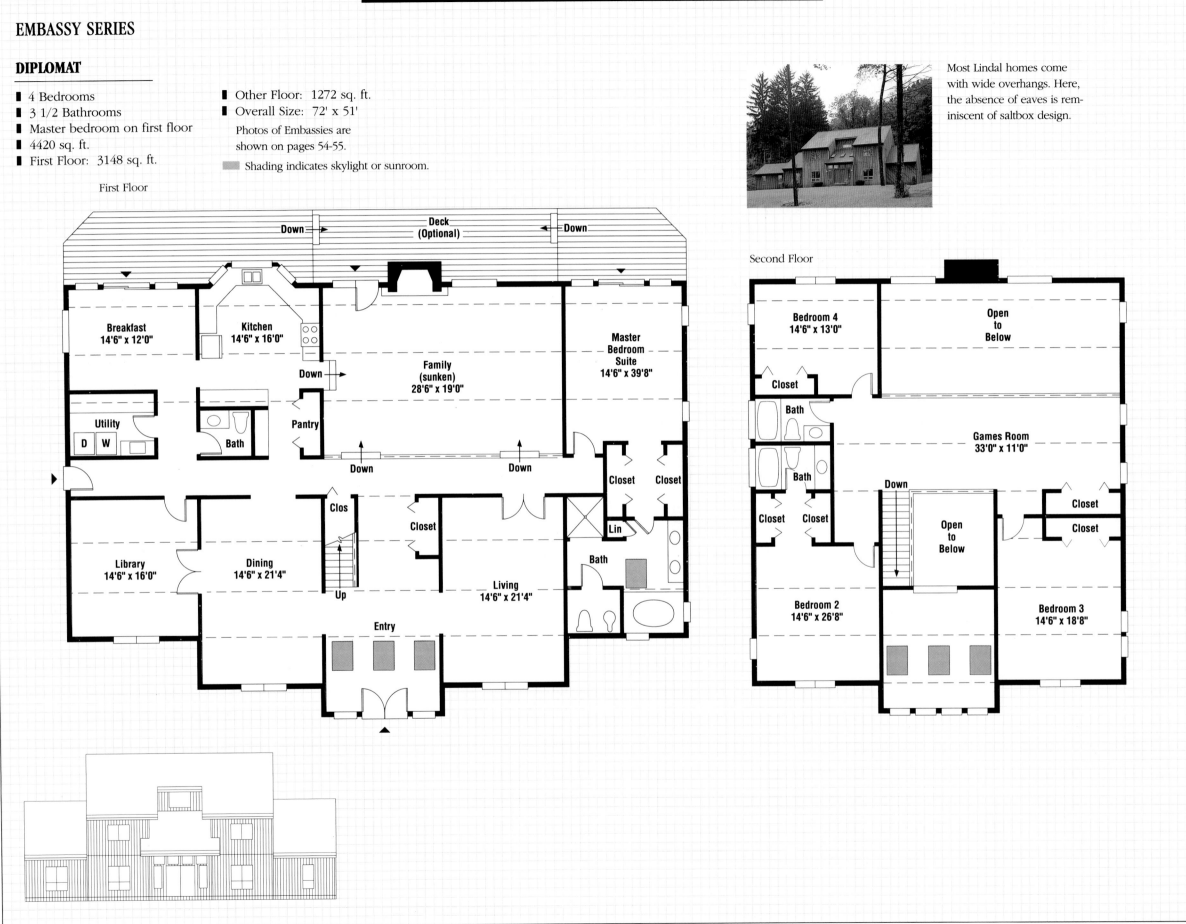

Second Floor

TOWN & COUNTRY SERIES

FAIRWAY

- 4+ Bedrooms
- 3 Bathrooms
- Master bedroom on second floor
- 2826 sq. ft.
- First Floor: 1461 sq. ft.
- Other Floor: 1365 sq. ft.
- Overall Size: 50' x 32'

Photos of Town & Countries are shown on pages 56-57.

This photo highlights the beauty of architectural grade glu-laminated beams and wood ceiling planks.

You can top your home with different roofs, among them:

1. Justus standard solid plank roof. R-38.

2. Lindal standard cavity roof. R-33.

3. Lindal Polar Cap 3 cavity roof. R-63. Optional.

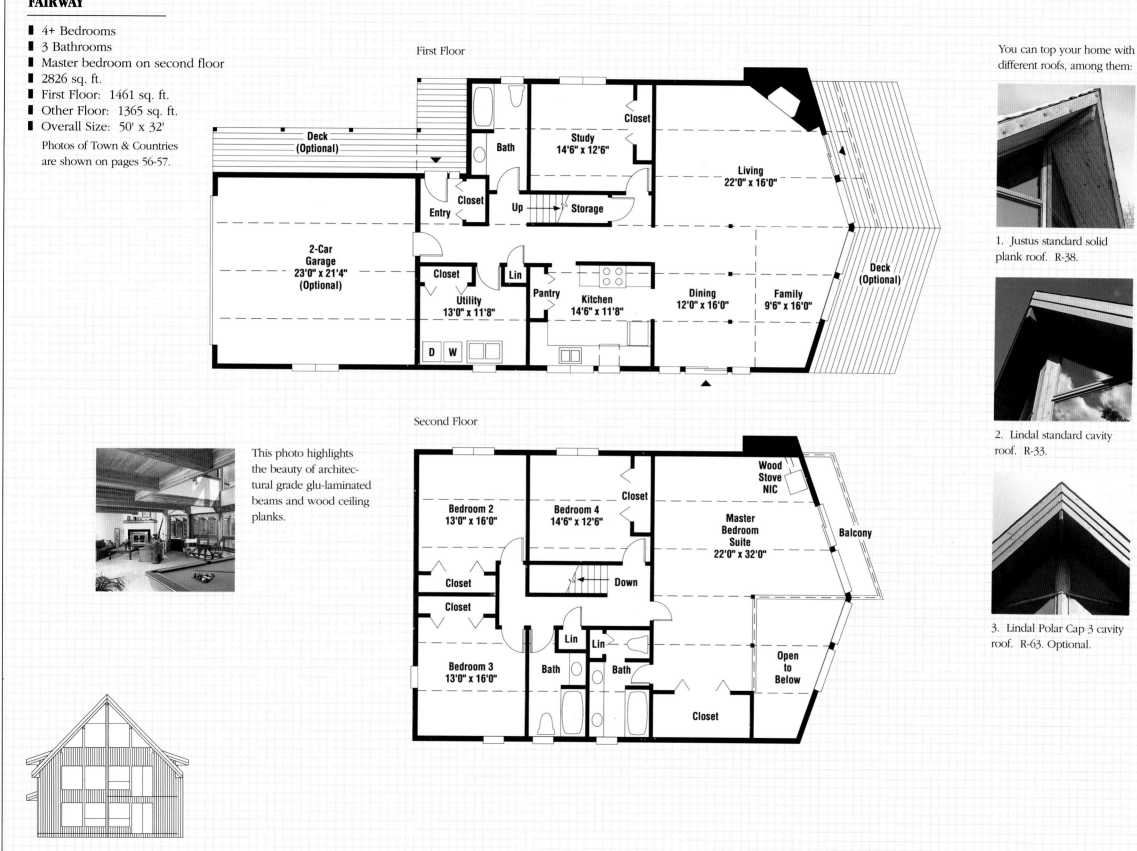

First Floor

Deck (Optional)

Bath

Study 14'6" x 12'6"

Closet

Living 22'0" x 16'0"

Closet

Entry

Up

Storage

2-Car Garage 23'0" x 21'4" (Optional)

Closet

Lin

Pantry

Kitchen 14'6" x 11'8"

Dining 12'0" x 16'0"

Family 9'6" x 16'0"

Deck (Optional)

Utility 13'0" x 11'8"

D   W

Second Floor

Bedroom 2 13'0" x 16'0"

Bedroom 4 14'6" x 12'6"

Closet

Wood Stove NIC

Master Bedroom Suite 22'0" x 32'0"

Balcony

Closet

Down

Closet

Lin   Lin

Bath   Bath

Bedroom 3 13'0" x 16'0"

Open to Below

Closet

TOWN & COUNTRY SERIES

FAIRLANE

- 4+ Bedrooms
- 3 Bathrooms
- Master bedroom on second floor
- 3388 sq. ft.
- First Floor: 1796 sq. ft.
- Other Floor: 1592 sq. ft.
- Overall Size: 51' x 37'

Photos of Town & Countries are shown on pages 56-57.

Shading indicates skylight or sunroom.

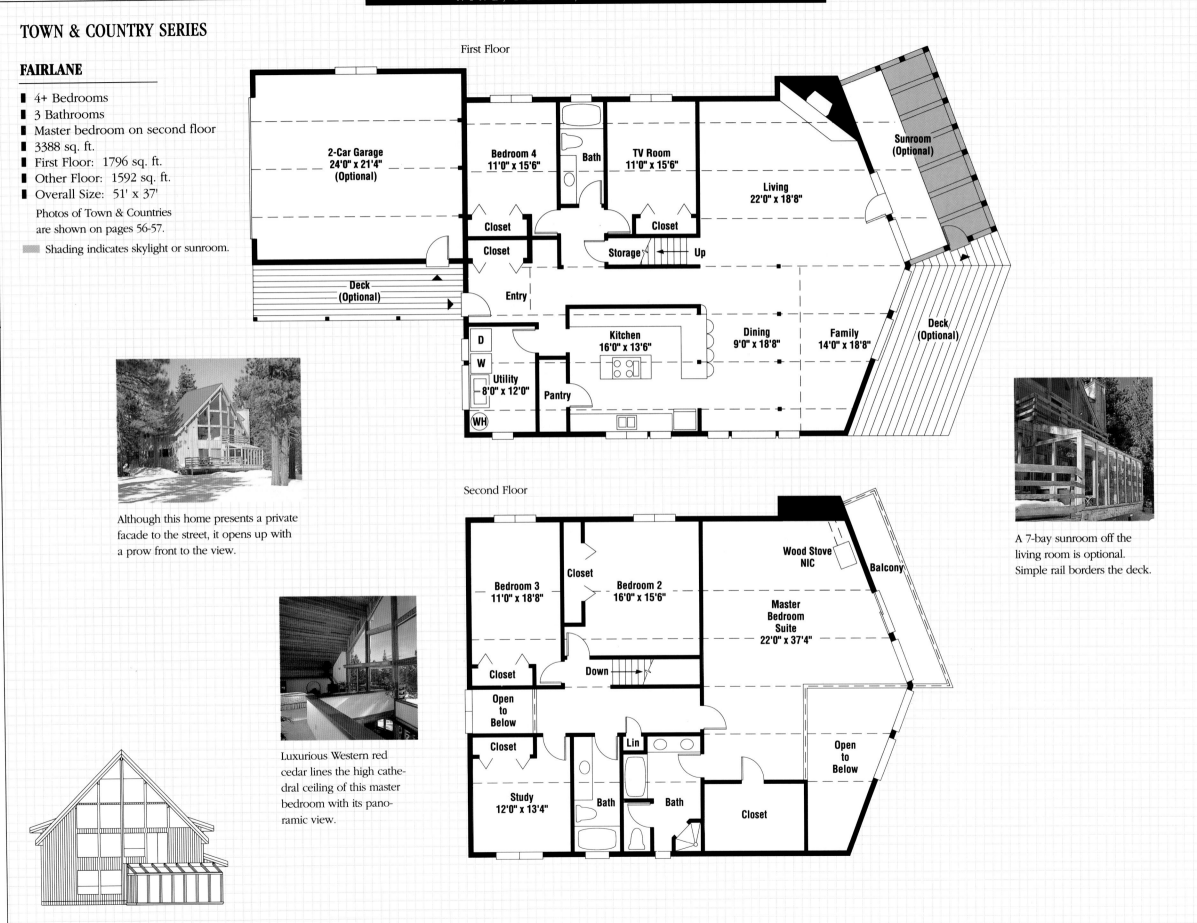

First Floor

2-Car Garage
24'0" x 21'4"
(Optional)

Bedroom 4
11'0" x 15'6"

Bath

TV Room
11'0" x 15'6"

Sunroom
(Optional)

Living
22'0" x 18'8"

Closet

Closet

Closet

Storage

Up

Deck
(Optional)

Entry

Deck
(Optional)

D

W

Kitchen
16'0" x 13'6"

Dining
9'0" x 18'8"

Family
14'0" x 18'8"

Utility
8'0" x 12'0"

Pantry

WH

Although this home presents a private facade to the street, it opens up with a prow front to the view.

A 7-bay sunroom off the living room is optional. Simple rail borders the deck.

Second Floor

Bedroom 3
11'0" x 18'8"

Closet

Bedroom 2
16'0" x 15'6"

Wood Stove
NIC

Balcony

Master
Bedroom
Suite
22'0" x 37'4"

Closet

Down

Open
to
Below

Closet

Open
to
Below

Lin

Study
12'0" x 13'4"

Bath

Bath

Closet

Luxurious Western red cedar lines the high cathedral ceiling of this master bedroom with its panoramic view.

CENTENNIAL SERIES

CENTURION

- 2 Bedrooms
- 2 1/2 Bathrooms
- Master bedroom on second floor
- 2172 sq. ft.
- First Floor: 1120 sq. ft.
- Other Floor: 1052 sq. ft.
- Overall Size: 49' x 27'

Photos of Centennials are shown on pages 58-59.

This detail shows the underside of the balcony; cantilevered glu-laminated beams support the weight. 2x6 cedar (crossing above the beams) provides the decking.

In high wind, snow or earthquake areas, some codes require larger mechanical connections between post and beam. This is an example of an optional decorative post to beam connector for high load areas.

First Floor

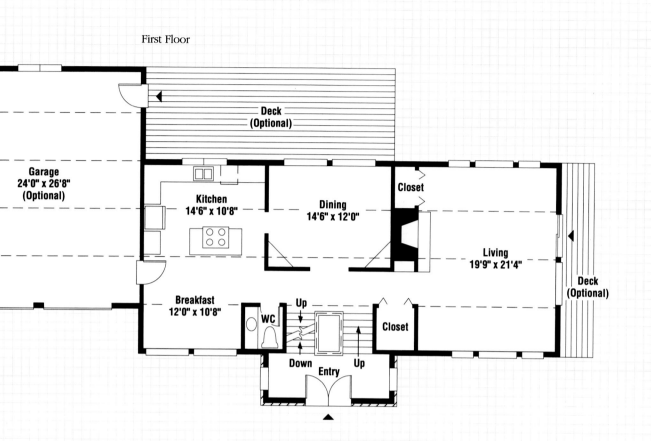

Garage
24'0" x 26'8"
(Optional)

Deck
(Optional)

Kitchen
14'6" x 10'8"

Dining
14'6" x 12'0"

Closet

Living
19'9" x 21'4"

Deck
(Optional)

Breakfast
12'0" x 10'8"

WC

Up

Closet

Down   Entry   Up

Second Floor

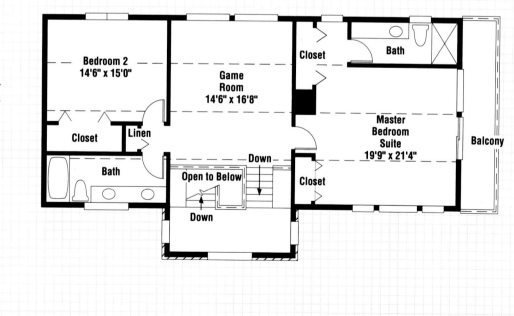

Bedroom 2
14'6" x 15'0"

Game
Room
14'6" x 16'8"

Closet

Bath

Closet   Linen

Master
Bedroom
Suite
19'9" x 21'4"

Balcony

Bath

Open to Below   Down   Closet

Down

Steel entry doors are for the truly energy-conscious, but raised panel cedar doors are an attractive option.

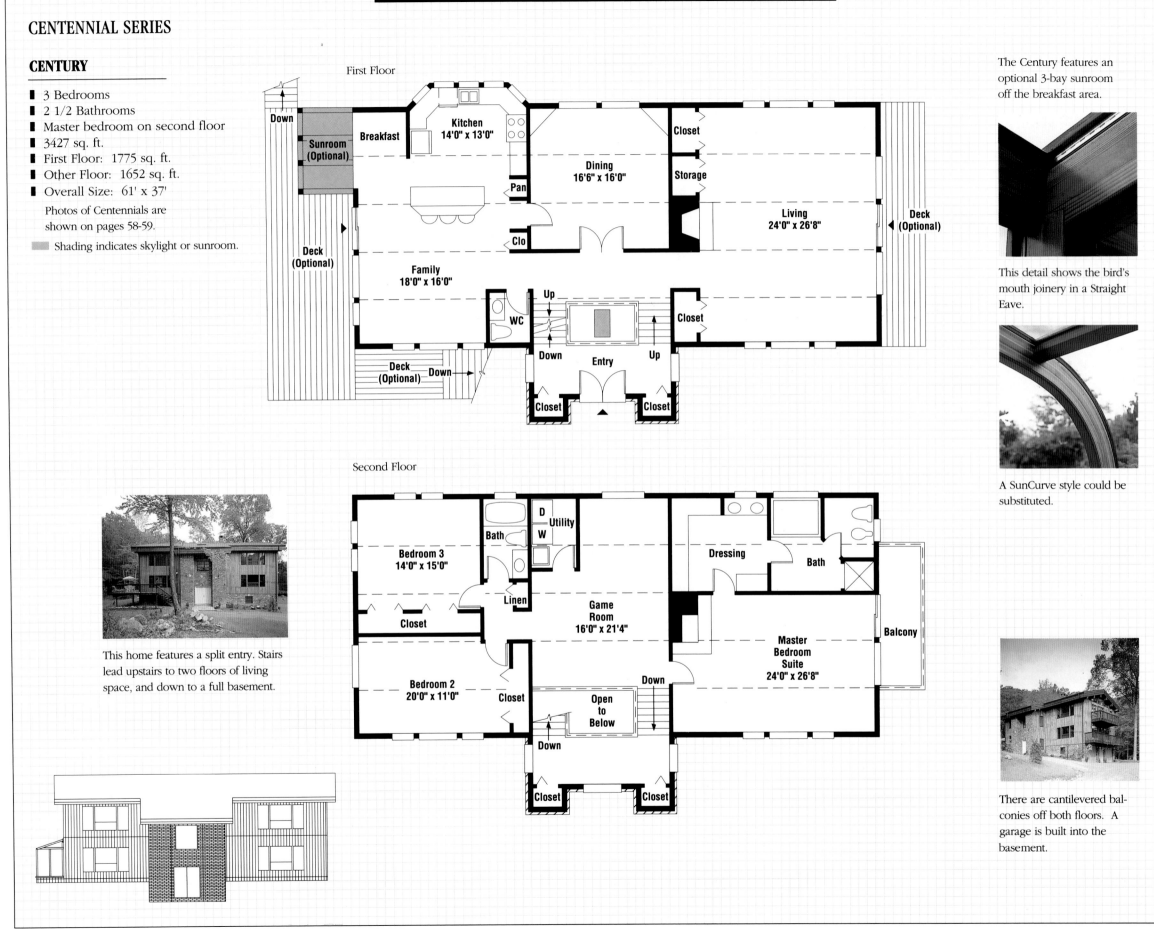

CENTENNIAL SERIES

CENTURY

- 3 Bedrooms
- 2 1/2 Bathrooms
- Master bedroom on second floor
- 3427 sq. ft.
- First Floor: 1775 sq. ft.
- Other Floor: 1652 sq. ft.
- Overall Size: 61' x 37'

Photos of Centennials are shown on pages 58-59.

Shading indicates skylight or sunroom.

First Floor

Second Floor

This home features a split entry. Stairs lead upstairs to two floors of living space, and down to a full basement.

The Century features an optional 3-bay sunroom off the breakfast area.

This detail shows the bird's mouth joinery in a Straight Eave.

A SunCurve style could be substituted.

There are cantilevered balconies off both floors. A garage is built into the basement.

MASTERPIECE SERIES

REMBRANDT

- 3 Bedrooms
- 2 1/2 Bathrooms
- Master bedroom on first floor
- 2107 sq. ft.
- First Floor: 1631 sq. ft.
- Other Floor: 476 sq. ft.
- Overall Size: 58' x 42'

Photos of Masterpieces are
shown on pages 60-61.

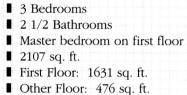

This detail shows a sliding
glass door opening from
the multi-sided pavilion to
the wide expanse of cedar
decking.

First Floor

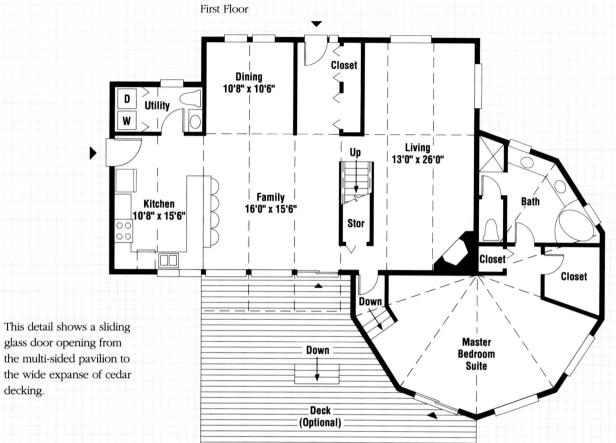

Dining
10'8" x 10'6"

Closet

D
W
Utility

Kitchen
10'8" x 15'6"

Family
16'0" x 15'6"

Up

Living
13'0" x 26'0"

Stor

Bath

Down

Down

Closet

Closet

Master
Bedroom
Suite

Deck
(Optional)

Inside the pavilion, the ceiling, with its many beams
meeting at the apex, is truly
spectacular.

Second Floor

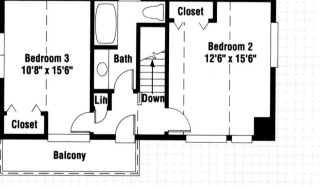

Bedroom 3
10'8" x 15'6"

Closet

Bath

Lin

Closet

Down

Closet

Bedroom 2
12'6" x 15'6"

Balcony

The Lindal specification
home comes with one
inch resawn cedar planks.
These interlocking planks
are tongued-and-grooved
for a tight fit.

MASTERPIECE SERIES

MICHELANGELO

- 4 Bedrooms
- 2 1/2 Bathrooms
- Master bedroom on first floor
- 3020 sq. ft.
- First Floor: 2148 sq. ft.
- Other Floor: 872 sq. ft.
- Overall Size: 69' x 46'

Photos of Masterpieces are shown on pages 60-61.

Shading indicates skylight or sunroom.

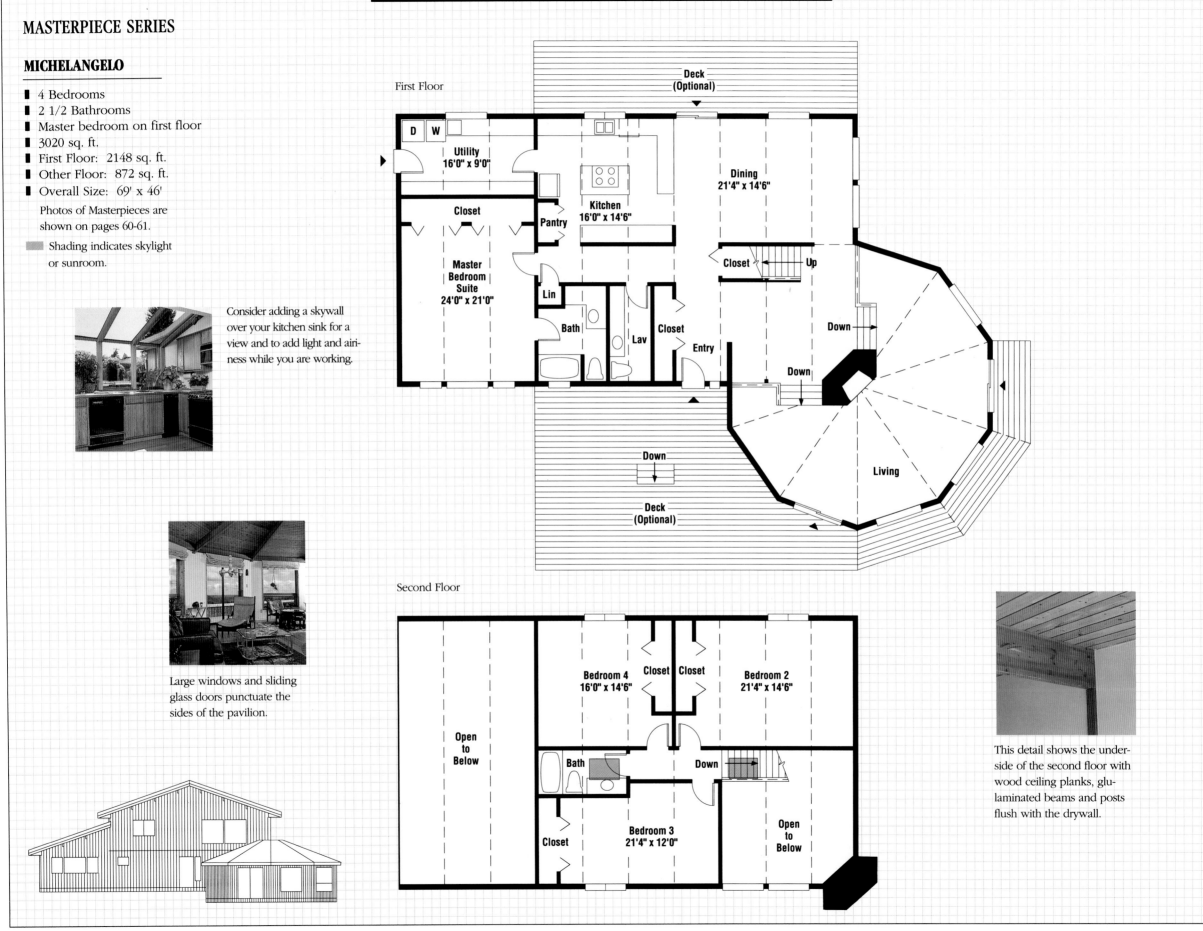

First Floor

Deck (Optional)

Utility 16'0" x 9'0"

Closet

Master Bedroom Suite 24'0" x 21'0"

Pantry

Kitchen 16'0" x 14'6"

Dining 21'4" x 14'6"

Closet

Up

Lin

Bath

Lav

Closet

Entry

Down

Down

Down

Living

Down

Deck (Optional)

Second Floor

Bedroom 4 16'0" x 14'6"

Closet **Closet**

Bedroom 2 21'4" x 14'6"

Open to Below

Bath

Down

Closet

Bedroom 3 21'4" x 12'0"

Open to Below

Consider adding a skywall over your kitchen sink for a view and to add light and airiness while you are working.

Large windows and sliding glass doors punctuate the sides of the pavilion.

This detail shows the underside of the second floor with wood ceiling planks, glu-laminated beams and posts flush with the drywall.

MASTERPIECE SERIES

MOZART

- 4 Bedrooms
- 2 1/2 Bathrooms
- Master bedroom on second floor
- 3703 sq. ft.
- First Floor: 2182 sq. ft.
- Other Floor: 1521 sq. ft.
- Overall Size: 73' x 49'

Photos of Masterpieces are shown on pages 60-61.

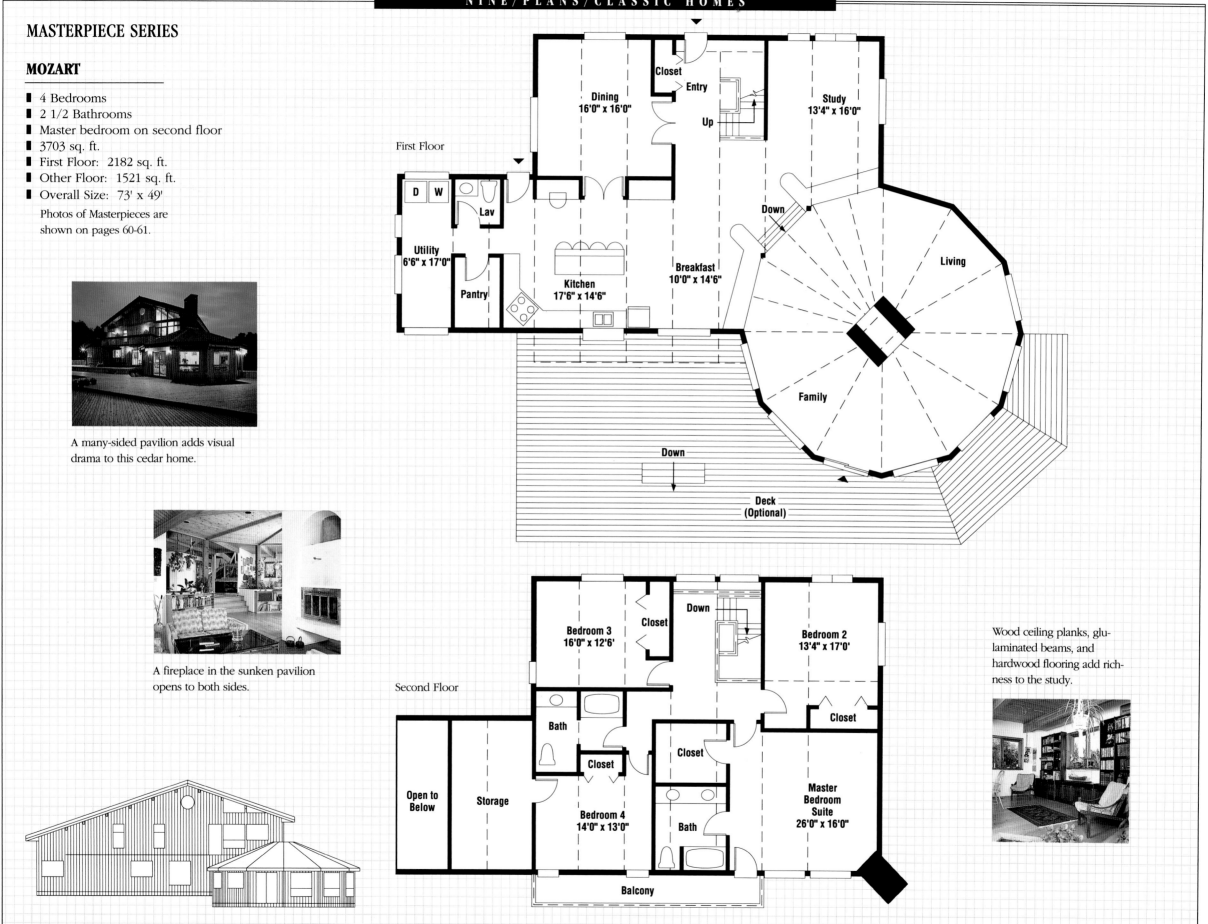

A many-sided pavilion adds visual drama to this cedar home.

A fireplace in the sunken pavilion opens to both sides.

Wood ceiling planks, glu-laminated beams, and hardwood flooring add richness to the study.

First Floor

Dining 16'0" x 16'0"
Closet
Entry
Up
Study 13'4" x 16'0"
D W
Lav
Down
Utility 6'6" x 17'0"
Breakfast 10'0" x 14'6"
Living
Pantry
Kitchen 17'6" x 14'6"
Family
Down
Deck (Optional)

Second Floor

Bedroom 3 16'0" x 12'6"
Closet
Down
Bedroom 2 13'4" x 17'0"
Bath
Closet
Open to Below
Storage
Closet
Closet
Bath
Bedroom 4 14'0" x 13'0"
Master Bedroom Suite 26'0" x 16'0"
Balcony

PROW STAR SERIES

BAY VISTA

- 1 Bedroom
- 2 Bathrooms
- Master bedroom on first floor
- 1387 sq. ft.
- First Floor: 1190 sq. ft.
- Other Floor: 197 sq. ft.
- Overall Size: 54' x 27'

Photos of Prow Stars are shown on pages 64-71.

Shading indicates skylight or sunroom.

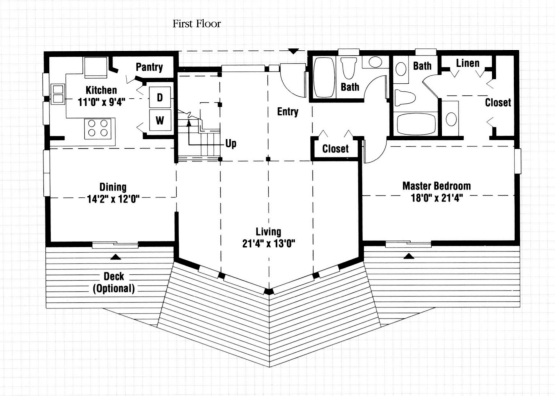

First Floor

Kitchen 11'0" x 9'4"
Pantry
D
W
Up
Entry
Bath
Bath
Linen
Closet
Closet
Dining 14'2" x 12'0"
Living 21'4" x 13'0"
Master Bedroom 18'0" x 21'4"
Deck (Optional)

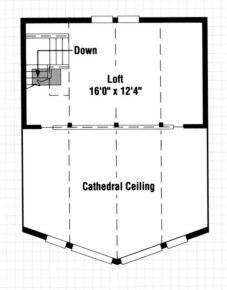

The spacious kitchen, open to the dining room, is one feature that makes the Bay Vista a favorite floor plan.

Second Floor

Down
Loft 16'0" x 12'4"
Cathedral Ceiling

The Bay Vista is a 4-module, 2-story prow with 1-story wings on both sides, and a rear entry. Here a detail of the 4"x 8" Justus solid timbers is shown.

Finely planed solid cedar timbers complement the Oriental decor of this home.

PROW STAR SERIES

NOVA VISTA

- 3 Bedrooms
- 2 Bathrooms
- Master bedroom on second floor
- 1747 sq. ft.
- First Floor: 1254 sq. ft.
- Other Floor: 493 sq. ft.
- Overall Size: 41' x 41'

Photos of Prow Stars are shown on pages 64-71.

▓ Shading indicates skylight or sunroom.

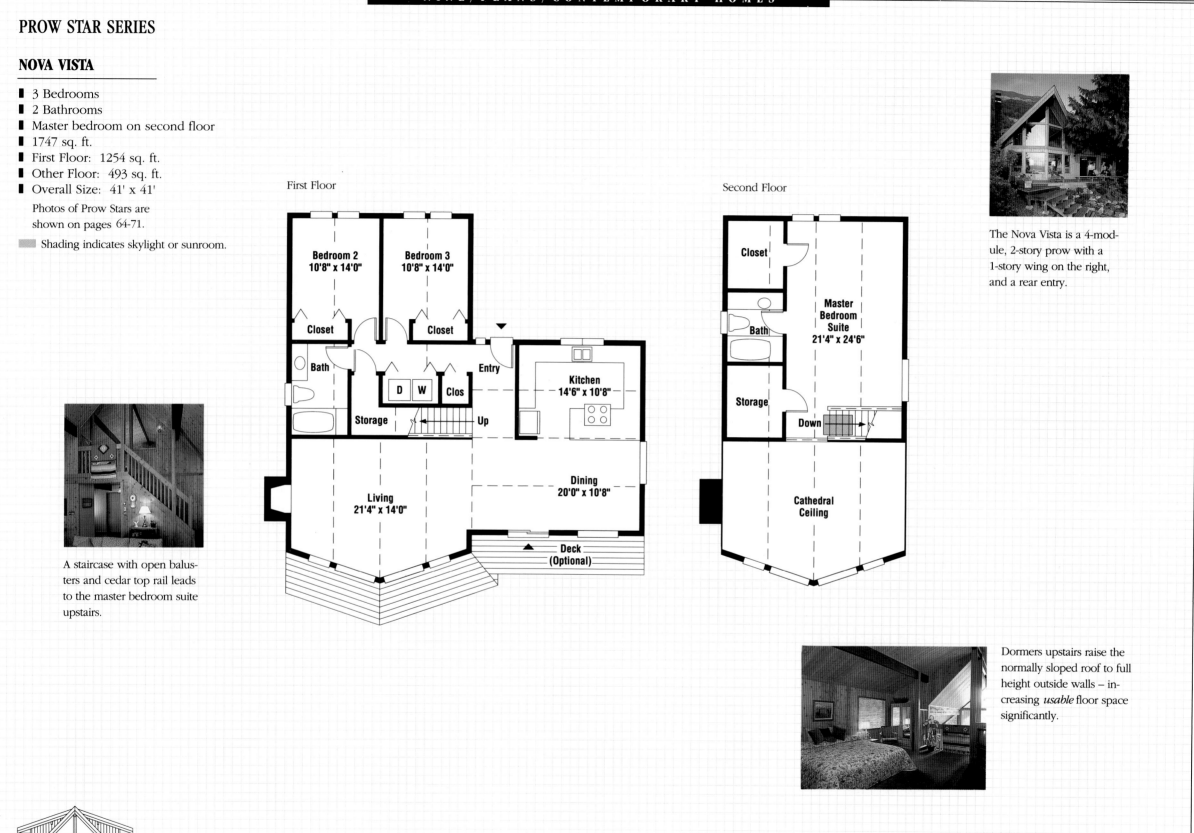

First Floor

Bedroom 2
10'8" x 14'0"

Bedroom 3
10'8" x 14'0"

Closet

Closet

Bath

Entry

D   W   Clos

Kitchen
14'6" x 10'8"

Storage

Up

Living
21'4" x 14'0"

Dining
20'0" x 10'8"

Deck
(Optional)

Second Floor

Closet

Bath

Master
Bedroom
Suite
21'4" x 24'6"

Storage

Down

Cathedral
Ceiling

A staircase with open balusters and cedar top rail leads to the master bedroom suite upstairs.

The Nova Vista is a 4-module, 2-story prow with a 1-story wing on the right, and a rear entry.

Dormers upstairs raise the normally sloped roof to full height outside walls – increasing *usable* floor space significantly.

PROW STAR SERIES

BUENA VISTA

- 3 Bedrooms
- 2 Bathrooms
- Master bedroom on second floor
- 1759 sq. ft.
- First Floor: 1322 sq. ft.
- Other Floor: 437 sq. ft.
- Overall Size: 50'x 35'

Photos of Prow Stars are shown on pages 64-71.

Shading indicates skylight or sunroom.

The Buena Vista is a 2-story prow with a 1-story wing canted on the right, and a side entry on the left.

First Floor

Bedroom 2
12'7" x 12'0"

Closet

Bedroom 3
11'8" x 12'0"

Deck
(Optional)

Closet   WH   W   Lin

Up   D   Bath

Closet

Living
26'8" x 16'0"

Family
18'7" x 10'8"

Kitchen
9'6"
x
10'8"

Dining
13'0" x 10'8"

Deck
(Optional)

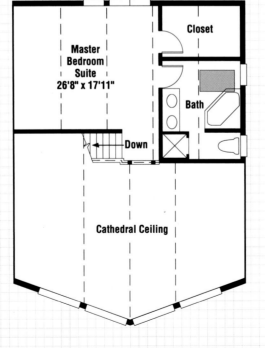

A prow front maximizes a view. Note how the drywall sets off the cedar accents.

Second Floor

Master
Bedroom
Suite
26'8" x 17'11"

Closet

Bath

Down

Cathedral Ceiling

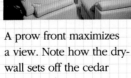

For bathrooms, Lindal provides moisture resistant "greenboard" rather than regular drywall around tub enclosures and behind toilets and vanities.

PROW STAR SERIES

CASCADE VISTA

- 2 Bedrooms
- 2 Bathrooms
- Master bedroom on first floor
- 1817 sq. ft.
- First Floor: 1549 sq. ft.
- Other Floor: 268 sq. ft.
- Overall Size: 59' x 32'

Photos of Prow Stars are shown on pages 64-71.

Shading indicates skylight or sunroom.

First Floor

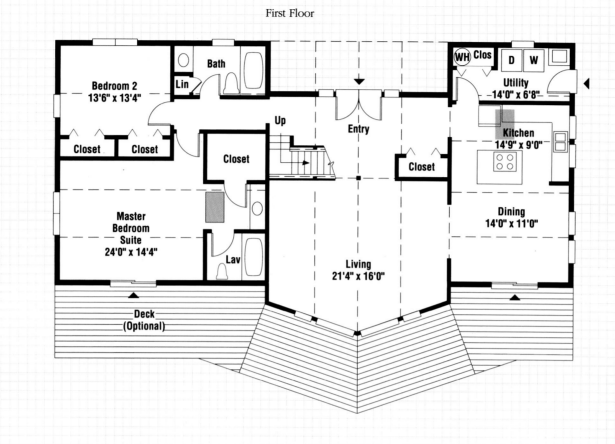

A skylight over the kitchen bathes the kitchen/dining area with light.

The Cascade Vista is a 4-module, 2-story prow, with 1-story wings on both sides and a recessed covered entry on the rear. Here it was built on a walkout daylight basement.

Second Floor

The windows of the prow frame the view. The deck configuration matches the prow.

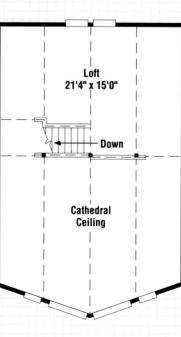

Stairs lead to the open loft above. The spacious foyer with its double doors adds grace to the home.

PROW STAR SERIES

SEA VISTA

- 3 Bedrooms
- 2 Bathrooms
- Master bedroom on second floor
- 1834 sq. ft.
- First Floor: 1453 sq. ft.
- Other Floor: 381 sq. ft.
- Overall Size: 60' x 36'

Photos of Prow Stars are shown on pages 64-71.

Shading indicates skylight or sunroom.

First Floor

Bedroom 2
17'1" x 10'8"

Closet

Family
10'8" x 16'0"

Kitchen
10'8" x 16'0"

Entry

Deck
(Optional)

Closet

Bedroom 3
15'7" x 10'8"

Lin   Pantry

Bath

W   D

Closet

Up

Dining
15'0" x 10'8"

Deck
(Optional)

Living
21'4' x 16'8"

Deck
(Optional)

Prows are a popular choice for framing panoramic views.

Second Floor

Balcony

Bath

Master
Bedroom
Suite
21'4" x 19'6"

Walk-in
Closet

Down

Cathedral
Ceiling

The Sea Vista is a 4-module, 2-story prow with 1-story wings swept back on both sides, and a rear entry on the right. Here's a detail of the wood-lined roof overhang on the prow.

This detail shows the 2x2 rails and 2x4 top rail in our all-cedar deck package.

PROW STAR SERIES

OMNI VISTA

- 3 Bedrooms
- 3 Bathrooms
- Master bedroom on second floor
- 2021 sq. ft.
- First Floor: 1672 sq. ft.
- Other Floor: 349 sq. ft.
- Overall Size: 59' x 36'

Photos of Prow Stars are shown on pages 64-71.

 Shading indicates skylight or sunroom.

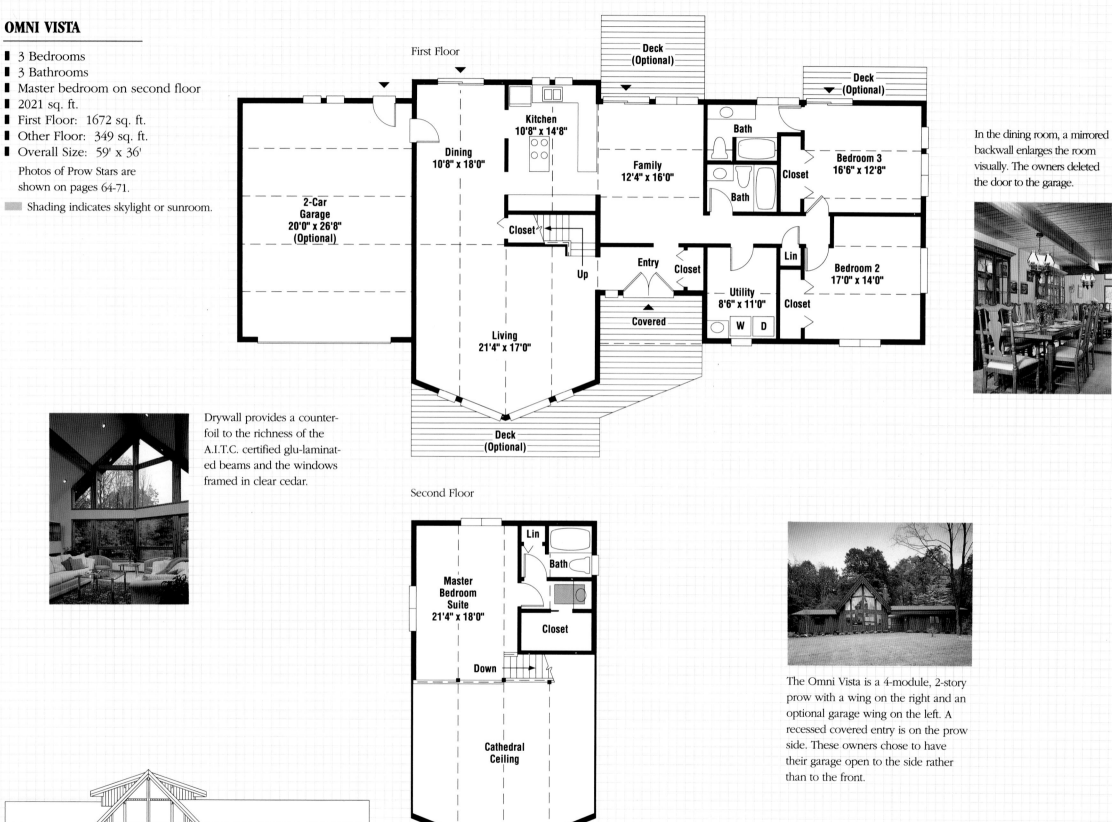

First Floor

Deck (Optional)

Deck (Optional)

Kitchen 10'8" x 14'8"

Dining 10'8" x 18'0"

Bath

Bedroom 3 16'6" x 12'8"

Closet

Family 12'4" x 16'0"

Bath

2-Car Garage 20'0" x 26'8" (Optional)

Closet

Up

Entry

Closet

Lin

Bedroom 2 17'0" x 14'0"

Utility 8'6" x 11'0"

Closet

Covered

W D

Living 21'4" x 17'0"

Deck (Optional)

Second Floor

Lin

Bath

Master Bedroom Suite 21'4" x 18'0"

Closet

Down

Cathedral Ceiling

In the dining room, a mirrored backwall enlarges the room visually. The owners deleted the door to the garage.

Drywall provides a counterfoil to the richness of the A.I.T.C. certified glu-laminated beams and the windows framed in clear cedar.

The Omni Vista is a 4-module, 2-story prow with a wing on the right and an optional garage wing on the left. A recessed covered entry is on the prow side. These owners chose to have their garage open to the side rather than to the front.

PROW STAR SERIES

LAKE VISTA

- 2 Bedrooms
- 2 Bathrooms
- Master bedroom on second floor
- 1976 sq. ft.
- First Floor: 1425 sq. ft.
- Other Floor: 551 sq. ft.
- Overall Size: 47' x 37'

Photos of Prow Stars are shown on pages 64-71.

Shading indicates skylight or sunroom.

The Lake Vista is a 4-module, 2-story prow with one wing on the right, and a rear entry. Here solid cedar timbers add warmth to the living room.

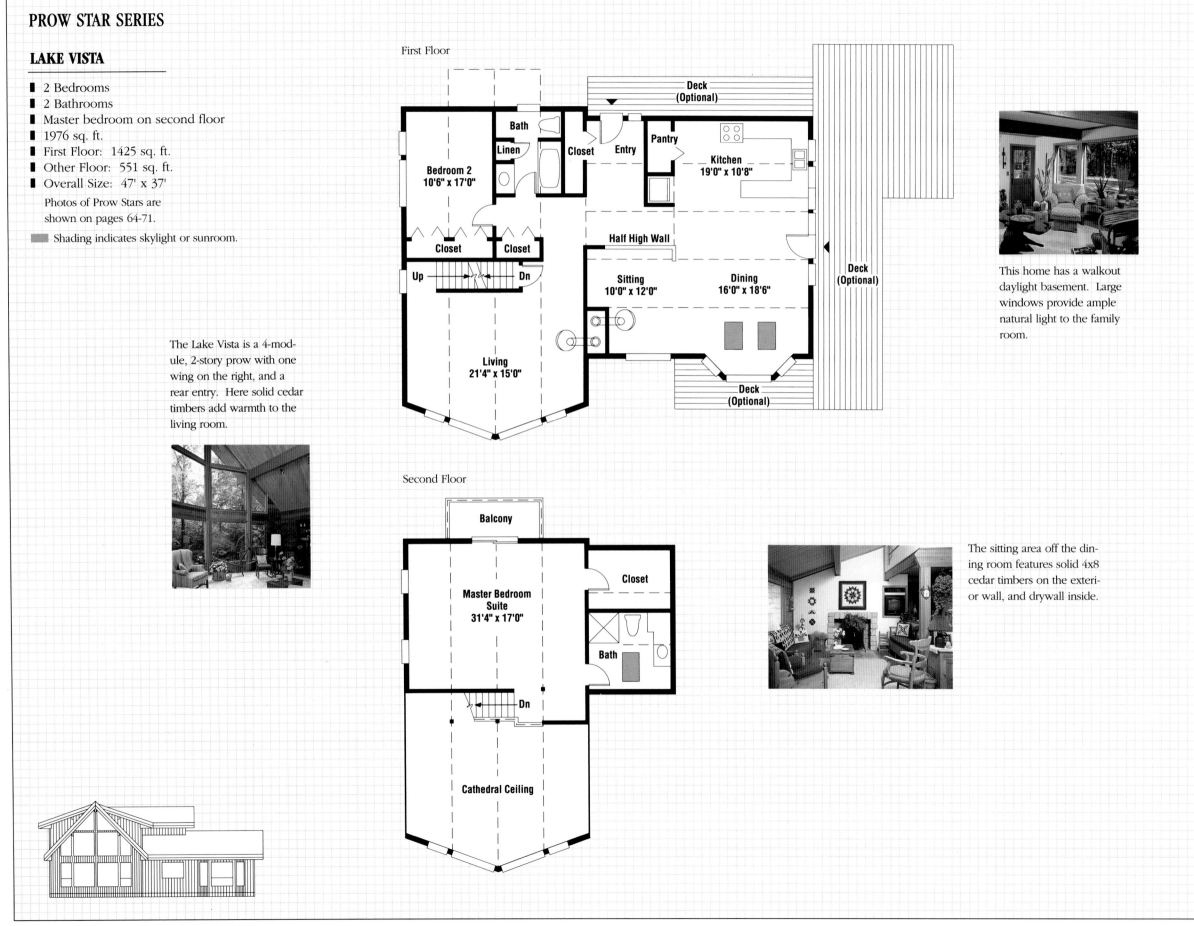

First Floor

Deck (Optional)

Bath

Linen

Closet

Entry

Pantry

Kitchen
19'0" x 10'8"

Bedroom 2
10'6" x 17'0"

Half High Wall

Closet   Closet

Up   Dn

Sitting
10'0" x 12'0"

Dining
16'0" x 18'6"

Deck (Optional)

Living
21'4" x 15'0"

Deck (Optional)

Second Floor

Balcony

Closet

Master Bedroom Suite
31'4" x 17'0"

Bath

Dn

Cathedral Ceiling

This home has a walkout daylight basement. Large windows provide ample natural light to the family room.

The sitting area off the dining room features solid 4x8 cedar timbers on the exterior wall, and drywall inside.

PRESIDENTIAL PROW STAR SERIES

LINCOLN

- 3 Bedrooms
- 2 1/2 Bathrooms
- Master bedroom on second floor
- 2047 sq. ft.
- First Floor: 1279 sq. ft.
- Other Floor: 768 sq. ft.
- Overall Size: 46' x 38'

Photos of Presidential Prow Stars are shown on pages 72-75.

Shading indicates skylight or sunroom.

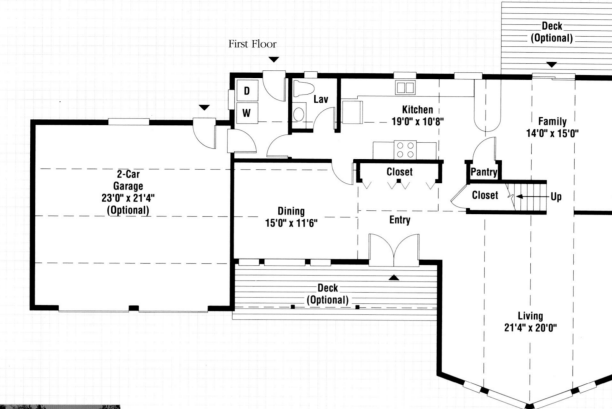

First Floor

Deck (Optional)

D
W
Lav
Kitchen 19'0" x 10'8"
Family 14'0" x 15'0"
Closet
Pantry
Closet
Up
2-Car Garage 23'0" x 21'4" (Optional)
Dining 15'0" x 11'6"
Entry
Living 21'4" x 20'0"
Deck (Optional)

In the living room, the owners ran their cedar liner diagonally, to form a herringbone pattern on the prow wall.

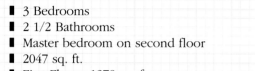

The Lincoln has a 4-module, 2-story prow with a 2-story wing on the left. The roof of the garage extends to its right, forming a covered porch over the deck and front entry.

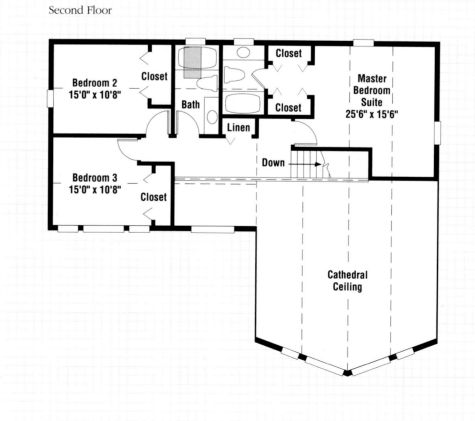

Second Floor

Bedroom 2 15'0" x 10'8"
Closet
Closet
Closet
Master Bedroom Suite 25'6" x 15'6"
Bath
Closet
Linen
Bedroom 3 15'0" x 10'8"
Closet
Down
Cathedral Ceiling

Glu-laminated beams, wood ceiling liner and cedar framed windows add rich notes in the dining room.

PRESIDENTIAL PROW STAR SERIES

JACKSON

- 4 Bedrooms
- 2 1/2 Bathrooms
- Master bedroom on second floor
- 2502 sq. ft.
- First Floor: 1470 sq. ft.
- Other Floor: 1032 sq. ft.
- Overall Size: 49' x 38'

Photos of Presidential Prow Stars are shown on pages 72-75.

Class A fiberglass shingles in a browntone to harmonize with the cedar are standard. Consider an opening skylight, too.

Thick butt, handsplit and resawn #1 cedar shakes are optional.

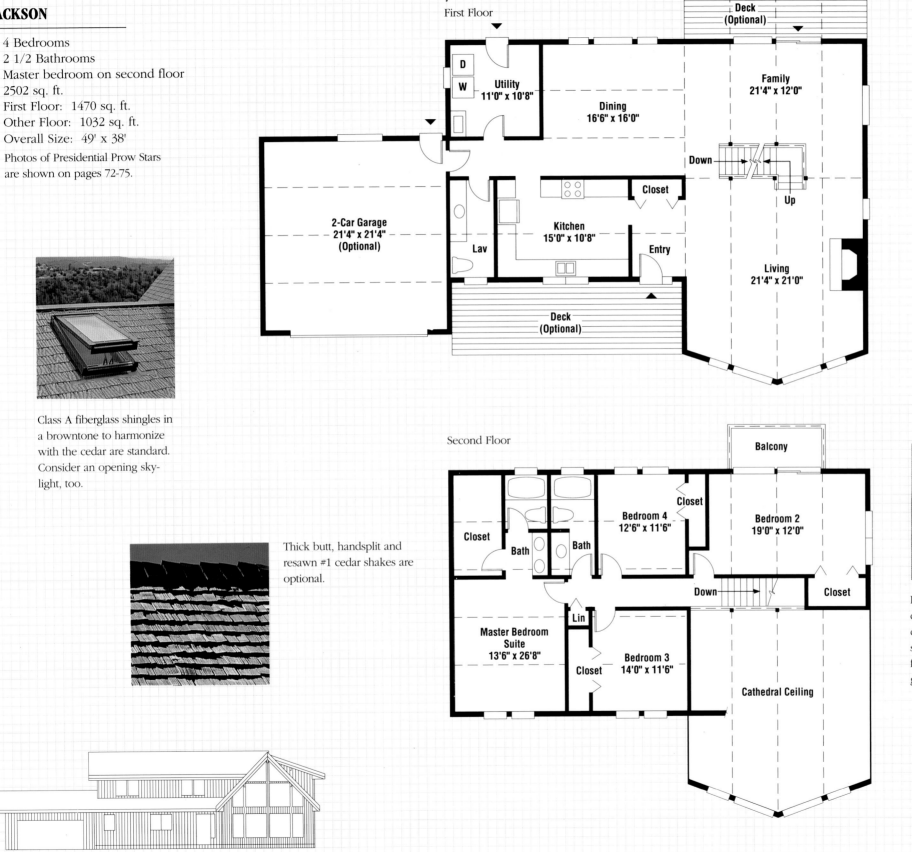

First Floor

Deck (Optional)

D
W
Utility 11'0" x 10'8"

Dining 16'6" x 16'0"

Family 21'4" x 12'0"

Down

Up

Closet

2-Car Garage 21'4" x 21'4" (Optional)

Kitchen 15'0" x 10'8"

Lav

Entry

Living 21'4" x 21'0"

Deck (Optional)

Second Floor

Balcony

Closet

Bedroom 4 12'6" x 11'6"

Bedroom 2 19'0" x 12'0"

Closet

Bath

Bath

Down

Closet

Lin

Master Bedroom Suite 13'6" x 26'8"

Bedroom 3 14'0" x 11'6"

Closet

Cathedral Ceiling

Beautiful 1'' cedar fascia and soffit trims the eaves.

Handsome raised panel clear cedar garage doors come in 8 ft., 9 ft. and 16 ft. sizes. They are pre-drilled for easy installation of garage door hardware.

CHALET STAR SERIES

GEMINI

- 3 Bedrooms
- 2 Bathrooms
- Master bedroom on second floor
- 1466 sq. ft.
- First Floor: 1162 sq. ft.
- Other Floor: 304 sq. ft.
- Overall Size: 45' x 31'

Photos of Chalet Stars are shown on pages 76-79.

Shading indicates skylight or sunroom.

The attractive open stairwell leading to the master bedroom suite on the second floor features optional glu-laminated stringers and treads with open risers.

First Floor

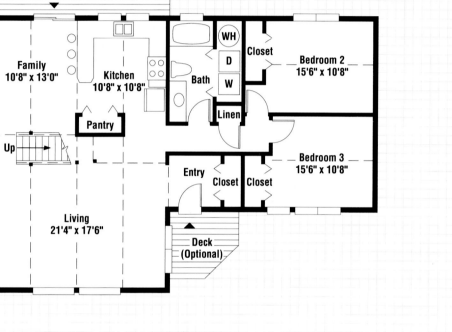

Deck (Optional)

Family 10'8" x 13'0"

Kitchen 10'8" x 10'8"

Pantry

Up

Living 21'4" x 17'6"

WH

D

W

Bath

Linen

Closet

Bedroom 2 15'6" x 10'8"

Bedroom 3 15'6" x 10'8"

Entry

Closet

Closet

Deck (Optional)

Second Floor

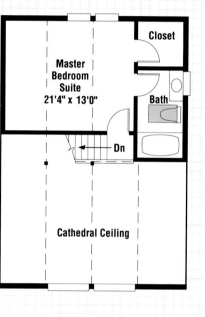

Master Bedroom Suite 21'4" x 13'0"

Closet

Bath

Dn

Cathedral Ceiling

2 1/4" cedar casing is provided for trimming doors and windows and for baseboard. Red oak is optional.

Exterior doors come standard with safety conscious lockset and deadbolt in an oil rubbed bronze finish (U.S.) and polished brass (Canada).

Consider a store door for certain entries. It comes pre-drilled for knob and deadbolt.

CHALET STAR SERIES

CAPRICORN

- 2 Bedrooms
- 2 Bathrooms
- Master bedroom on second floor
- 1483 sq. ft.
- First Floor: 1104 sq. ft.
- Other Floor: 379 sq. ft.
- Overall Size: 40' x 31'

Photos of Chalet Stars are shown on pages 76-79.

Shading indicates skylight or sunroom.

The owners added an outstanding fieldstone fireplace in the first two modules of the 5-module chalet.

First Floor

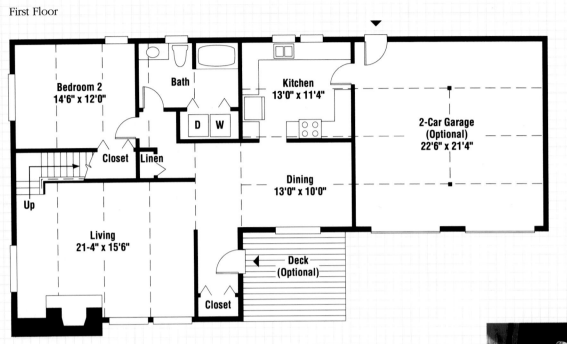

Bedroom 2 14'6" x 12'0"

Bath

Kitchen 13'0" x 11'4"

2-Car Garage (Optional) 22'6" x 21'4"

D W

Closet **Linen**

Dining 13'0" x 10'0"

Up

Living 21-4" x 15'6"

Deck (Optional)

Closet

Second Floor

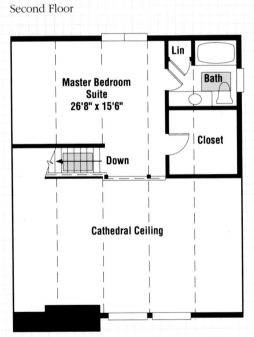

Master Bedroom Suite 26'8" x 15'6"

Lin

Bath

Down

Closet

Cathedral Ceiling

This detail shows the opening adjustable tension scissors hardware in an awning window.

This detail shows the rotogear crank handle on the window.

CHALET STAR SERIES

AQUARIUS

- 3 Bedrooms
- 2 Bathrooms
- Master bedroom on second floor
- 1670 sq. ft.
- First Floor: 1287 sq. ft.
- Other Floor: 383 sq. ft.
- Overall Size: 51' x 30'

Photos of Chalet Stars are shown on pages 76-79.

Shading indicates skylight or sunroom.

To achieve a more open feeling, the owners replaced the wall between the kitchen and dining room with suspended cabinets.

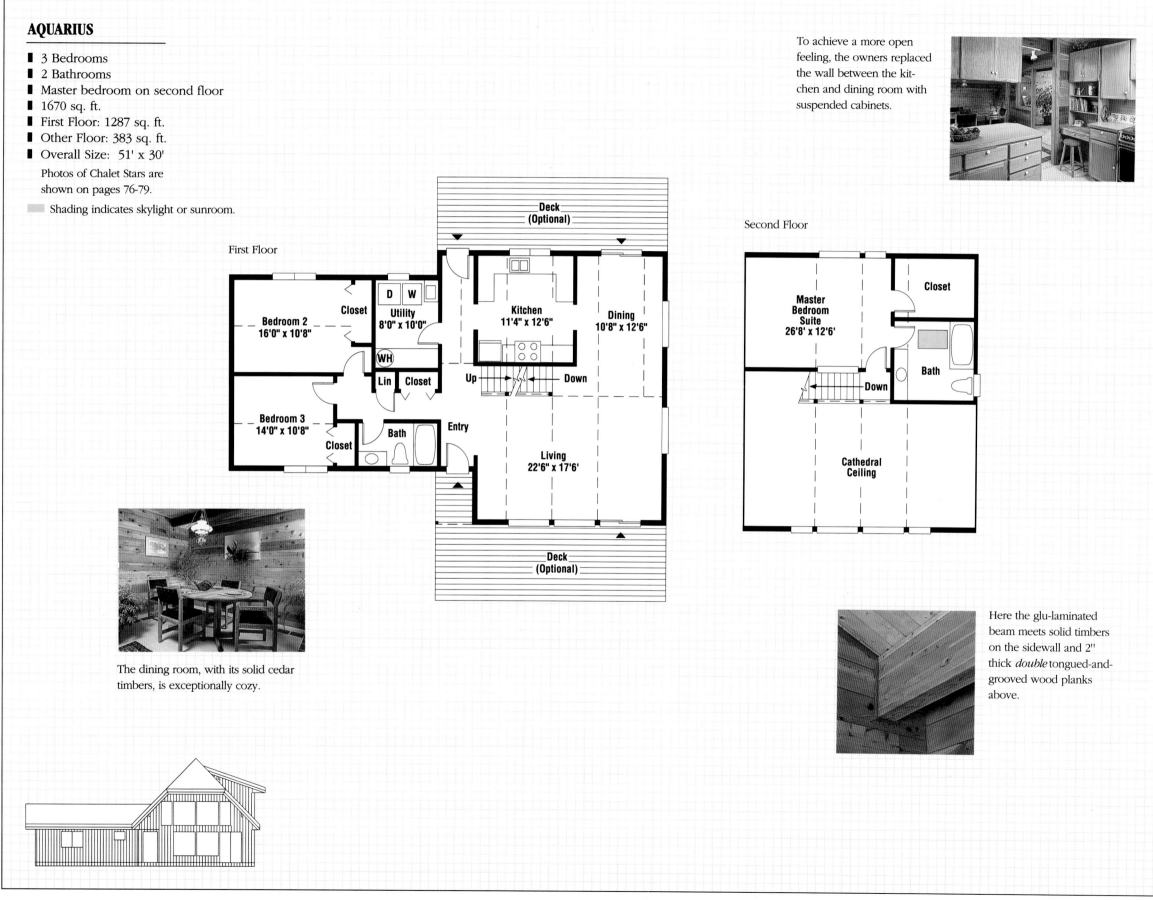

First Floor

Deck (Optional)

Bedroom 2 16'0" x 10'8"

Closet

D W

Utility 8'0" x 10'0"

WH

Lin   Closet

Bedroom 3 14'0" x 10'8"

Closet

Bath

Entry

Kitchen 11'4" x 12'6"

Dining 10'8" x 12'6"

Up   Down

Living 22'6" x 17'6"

Deck (Optional)

Second Floor

Master Bedroom Suite 26'8" x 12'6'

Closet

Bath

Down

Cathedral Ceiling

The dining room, with its solid cedar timbers, is exceptionally cozy.

Here the glu-laminated beam meets solid timbers on the sidewall and 2" thick *double* tongued-and-grooved wood planks above.

CHALET STAR SERIES

VIRGO

- 2 Bedrooms
- 1 3/4 Bathrooms
- Master bedroom on first floor
- 1676 sq. ft.
- First Floor: 1359 sq. ft.
- Other Floor 317 sq. ft.
- Overall Size: 48' x 32'

Photos of Chalet Stars are shown on pages 76-79.

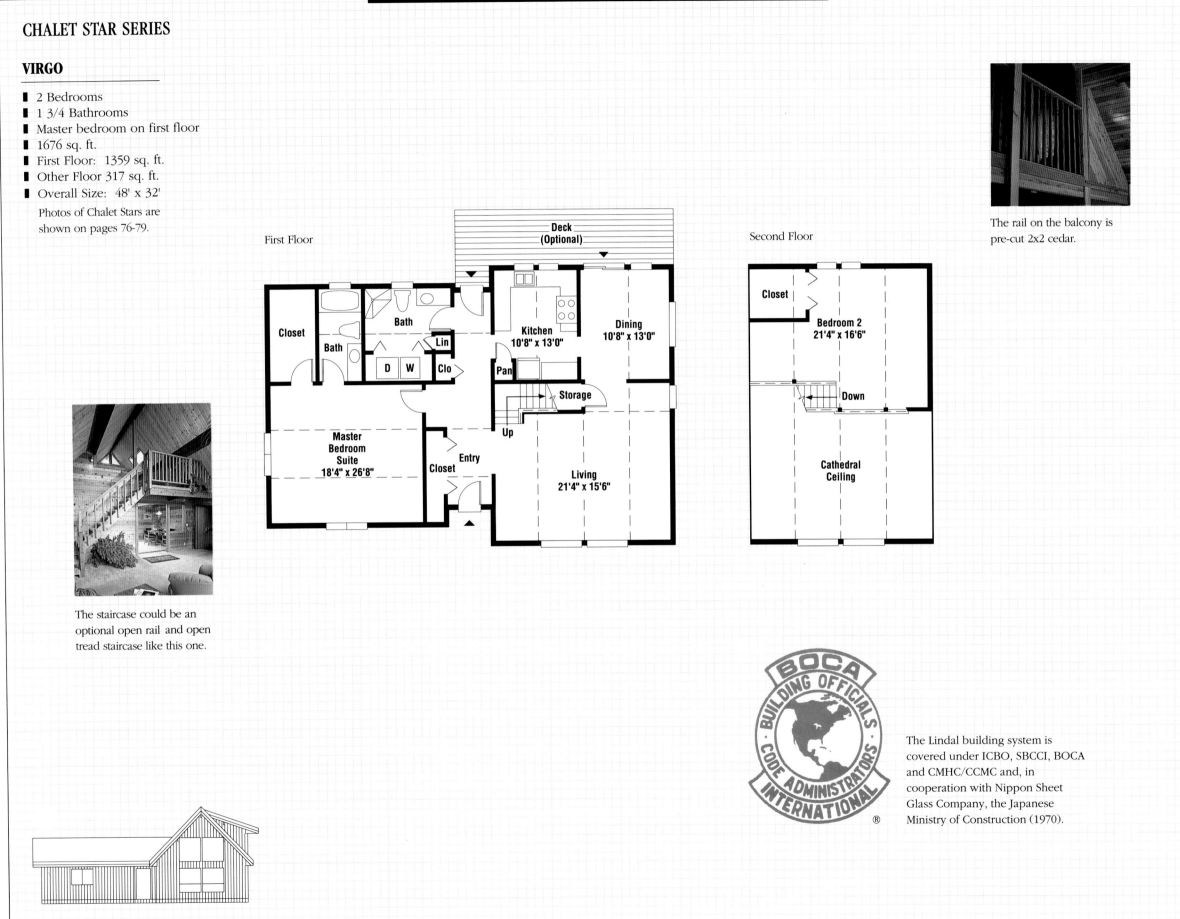

First Floor

Second Floor

The rail on the balcony is pre-cut 2x2 cedar.

The staircase could be an optional open rail and open tread staircase like this one.

The Lindal building system is covered under ICBO, SBCCI, BOCA and CMHC/CCMC and, in cooperation with Nippon Sheet Glass Company, the Japanese Ministry of Construction (1970).

CHALET STAR SERIES

LIBRA

- 3 Bedrooms
- 2 Bathrooms
- Master bedroom on first floor
- 1841 sq. ft.
- First Floor: 1507 sq. ft.
- Other Floor: 334 sq. ft.
- Overall Size: 59' x 34'

Photos of Chalet Stars are shown on pages 76-79.

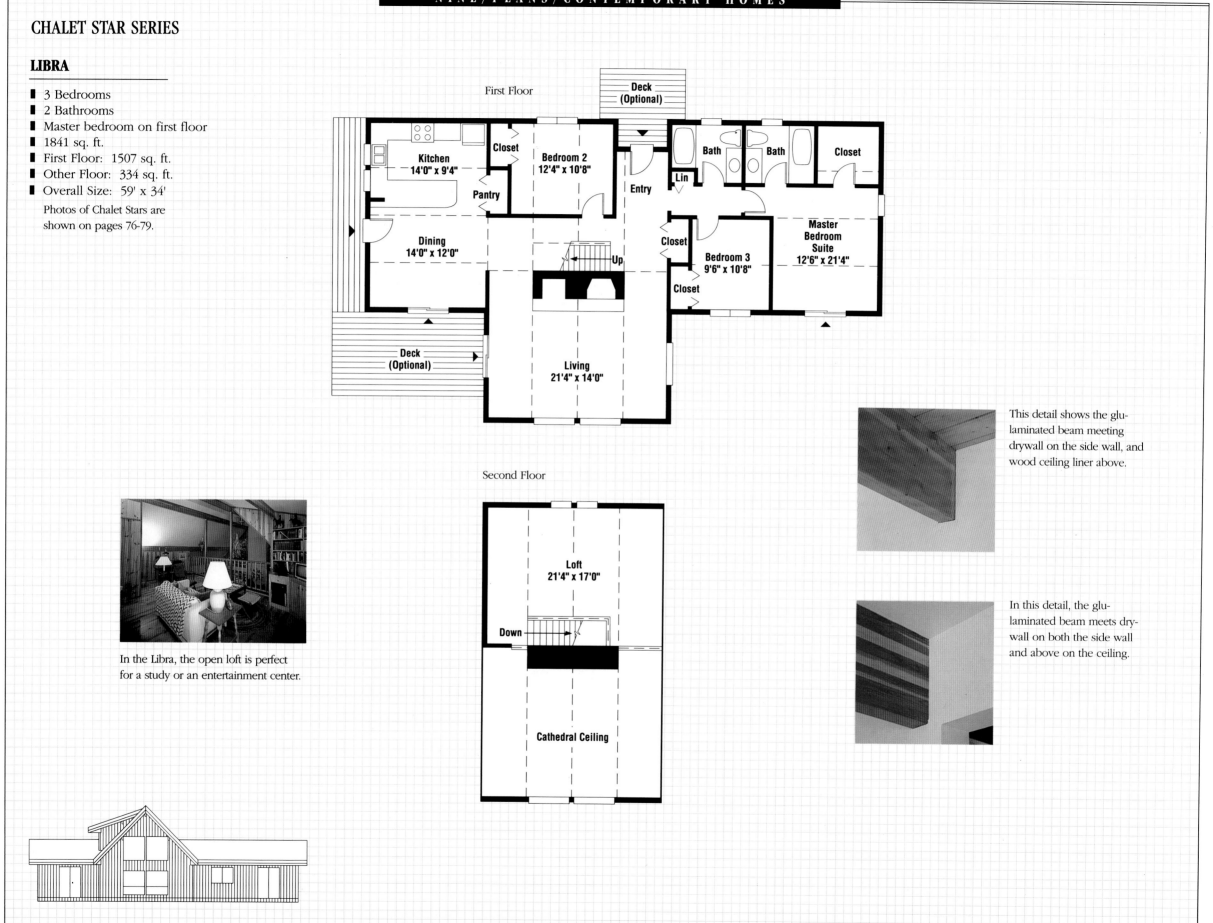

First Floor

Deck (Optional)

Kitchen 14'0" x 9'4"

Closet

Bedroom 2 12'4" x 10'8"

Bath | Bath

Closet

Pantry

Entry

Lin

Master Bedroom Suite 12'6" x 21'4"

Dining 14'0" x 12'0"

Up

Closet

Bedroom 3 9'6" x 10'8"

Closet

Deck (Optional)

Living 21'4" x 14'0"

Second Floor

Loft 21'4" x 17'0"

Down

Cathedral Ceiling

In the Libra, the open loft is perfect for a study or an entertainment center.

This detail shows the glu-laminated beam meeting drywall on the side wall, and wood ceiling liner above.

In this detail, the glu-laminated beam meets dry-wall on both the side wall and above on the ceiling.

CHALET STAR SERIES

ZODIAC

- 3 Bedrooms
- 2 1/2 Bathrooms
- Master bedroom on second floor
- 2255 sq. ft.
- First Floor: 1303 sq. ft.
- Other Floor: 952 sq. ft.
- Overall Size: 48' x 36'

Photos of Chalet Stars are
shown on pages 76-79.

In the Zodiac, two chalets intersect
at right angles. The chalet on the left
has a full second floor, whereas the
one on the right crops the second
floor to a loft.

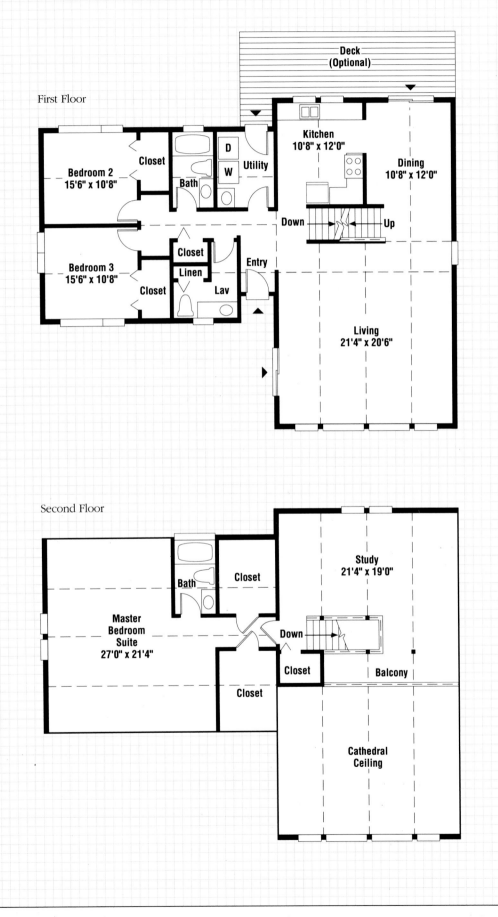

First Floor

Deck
(Optional)

Kitchen
10'8" x 12'0"

Dining
10'8" x 12'0"

Closet

D
W
Utility

Bath

Down   Up

Bedroom 2
15'6" x 10'8"

Closet
Linen

Entry

Bedroom 3
15'6" x 10'8"

Closet   Lav

Living
21'4" x 20'6"

Second Floor

Bath

Closet

Study
21'4" x 19'0"

Master
Bedroom
Suite
27'0" x 21'4"

Down

Closet   Balcony

Closet

Cathedral
Ceiling

Hardwood flooring is a
delightful addition to any
home. Here, Rustic Oak
with its lively coloration is
shown.

This photo shows a detail
of the corner of a solid wood
wall. Notice the *double*
tongues-and-grooves on the
ends of the timbers on the
right.

CONTEMPO PROW STAR SERIES

MERCURY

- 3 Bedrooms
- 2 Bathrooms
- 1515 sq. ft.
- Overall Size: 54' x 35'

Photos of Contempo Prow Stars are shown on pages 80-83.

The Contempo Prow Stars have a lower pitched roof line than the preceding Prow Stars and Chalet Stars. Here the Mercury has a 1-story wing with a front entry on the left.

The prow captures the view. Swinging patio doors open to the deck, and awning windows open for a breeze.

Cedar enthusiasts, these owners lined the home in multi hued cedar. They also added a low loft – accessible only by ladder – to the back of the prow.

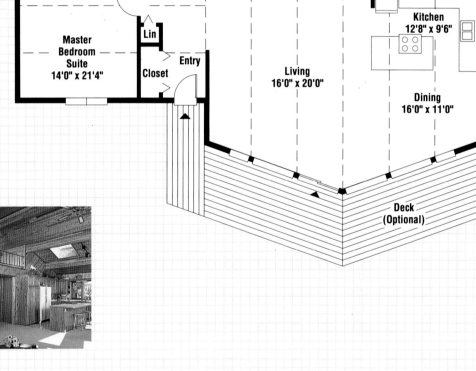

These owners added a striking stone fireplace in the living room. Beyond is the bedroom wing.

CONTEMPO PROW STAR SERIES

JUPITER

- 2 Bedrooms
- 2 Bathrooms
- 1573 sq. ft.
- Overall Size: 62' x 40'

Photos of Contempo Prow Stars are shown on pages 80-83.

▦ Shading indicates skylight or sunroom.

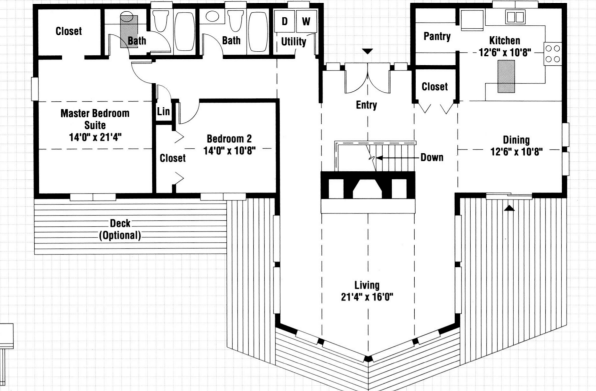

In the Jupiter, the 4/12 prow projects forward prominently from the two side wings. The formal entry, with optional raised panel cedar doors, is at the rear or street side.

VENUS

- 3 Bedrooms
- 2 Bathrooms
- 1593 sq. ft.
- Overall Size: 57' x 33'

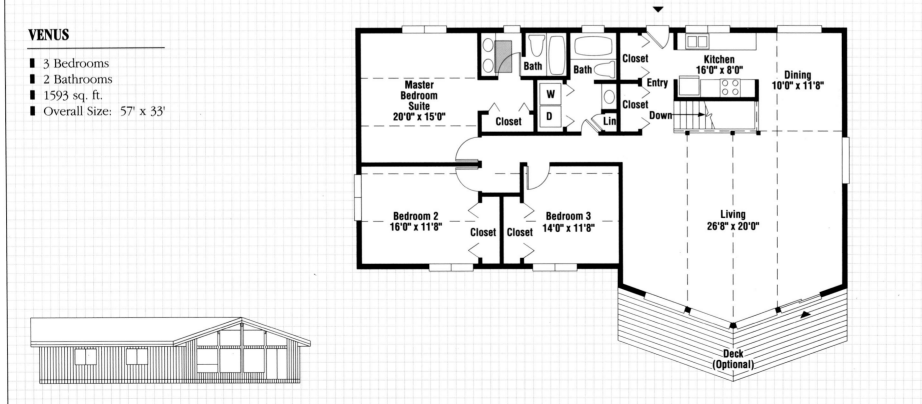

The owners built their Venus over a walkout daylight basement; the entry is at the rear.

CONTEMPO PROW STAR SERIES

GALAXY

- 3 Bedrooms
- 2 1/2 Bathrooms
- Master bedroom on first floor
- 3474 sq. ft.
- First Floor: 1719 sq. ft.
- Other Floor: 1755 sq. ft.
- Overall Size: 74' x 41'

Photos of Contempo Prow Stars
are shown on pages 80-83.

Shading indicates skylight or sunroom.

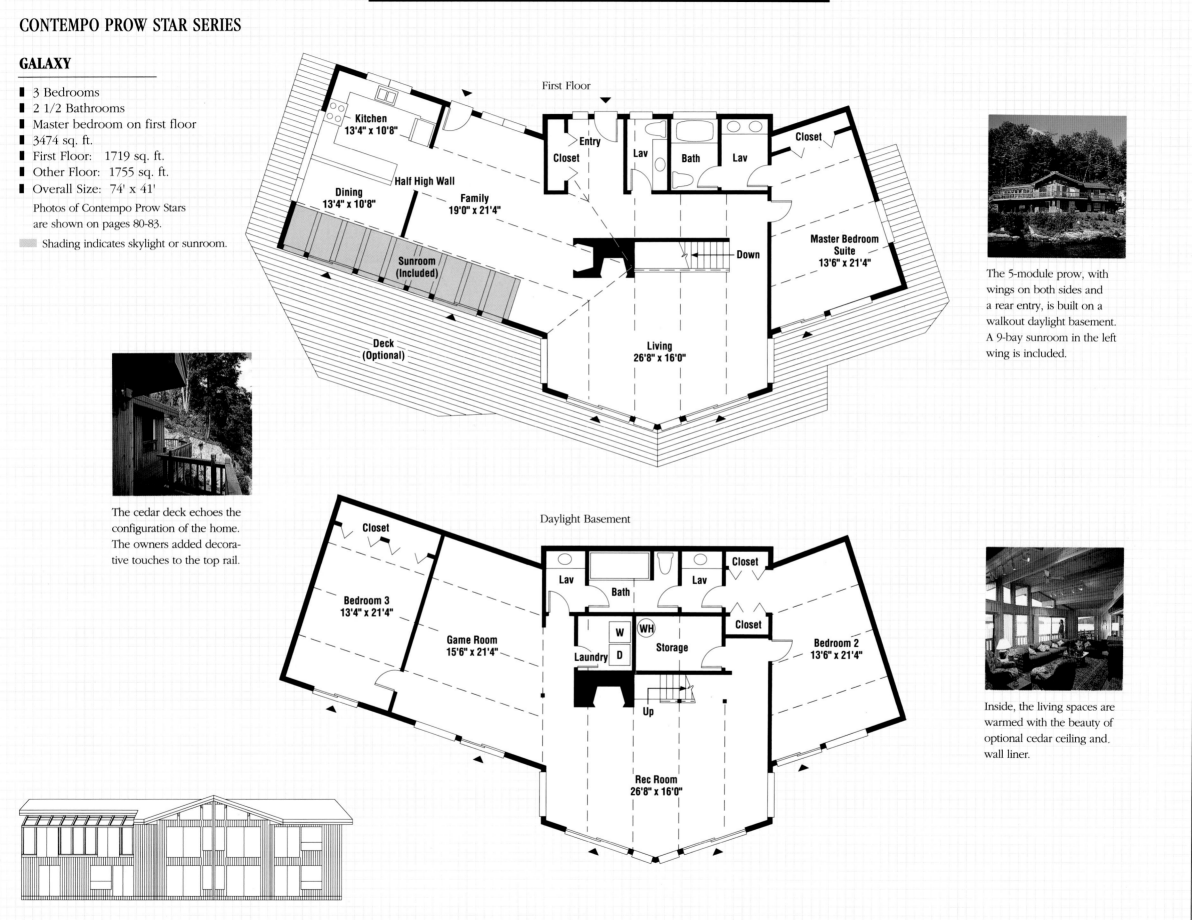

First Floor

Kitchen 13'4" x 10'8"

Entry

Closet

Lav

Bath

Lav

Closet

Dining 13'4" x 10'8"

Half High Wall

Family 19'0" x 21'4"

Master Bedroom Suite 13'6" x 21'4"

Sunroom (Included)

Down

Deck (Optional)

Living 26'8" x 16'0"

The cedar deck echoes the configuration of the home. The owners added decorative touches to the top rail.

The 5-module prow, with wings on both sides and a rear entry, is built on a walkout daylight basement. A 9-bay sunroom in the left wing is included.

Daylight Basement

Closet

Bedroom 3 13'4" x 21'4"

Game Room 15'6" x 21'4"

Lav

Bath

Lav

Closet

Closet

W WH

D Storage

Laundry

Bedroom 2 13'6" x 21'4"

Up

Rec Room 26'8" x 16'0"

Inside, the living spaces are warmed with the beauty of optional cedar ceiling and wall liner.

TRADEWINDS SERIES

WINDJAMMER

- 2 Bedrooms
- 2 Bathrooms
- Master bedroom on first floor
- 2042 sq. ft.
- First Floor: 1676 sq. ft.
- Other Floor: 366 sq. ft.
- Overall Size: 58' x 50'

Photos of Tradewinds are shown on pages 84-85.

Shading indicates skylight or sunroom.

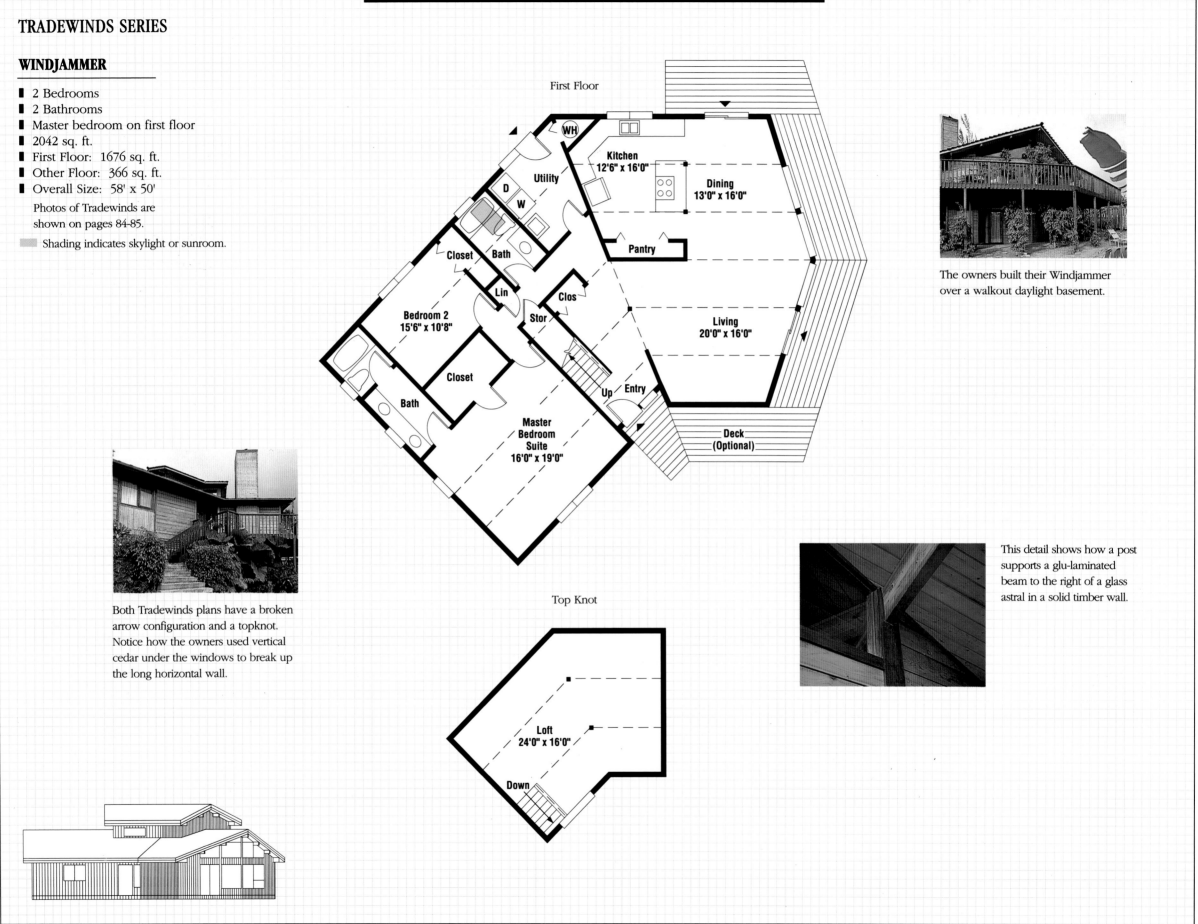

First Floor

WH

Kitchen
12'6" x 16'0"

Dining
13'0" x 16'0"

Utility

D

W

Pantry

Closet

Bath

Lin

Clos

Stor

Bedroom 2
15'6" x 10'8"

Living
20'0" x 16'0"

Closet

Bath

Up   Entry

Master
Bedroom
Suite
16'0" x 19'0"

Deck
(Optional)

The owners built their Windjammer over a walkout daylight basement.

Both Tradewinds plans have a broken arrow configuration and a topknot. Notice how the owners used vertical cedar under the windows to break up the long horizontal wall.

Top Knot

Loft
24'0" x 16'0"

Down

This detail shows how a post supports a glu-laminated beam to the right of a glass astral in a solid timber wall.

TRADEWINDS SERIES

WINDSONG

- 3 Bedrooms
- 2 Bathrooms
- Master bedroom on first floor
- 2283 sq. ft.
- First Floor: 1928 sq. ft.
- Other Floor: 355 sq. ft.
- Overall Size: 64' x 54'

Photos of Tradewinds are shown on pages 84-85.

▨ Shading indicates skylight or sunroom.

This kitchen observes the rule of the classic triangle: Sink, refrigerator and cooktop form a work triangle not exceeding 22 feet.

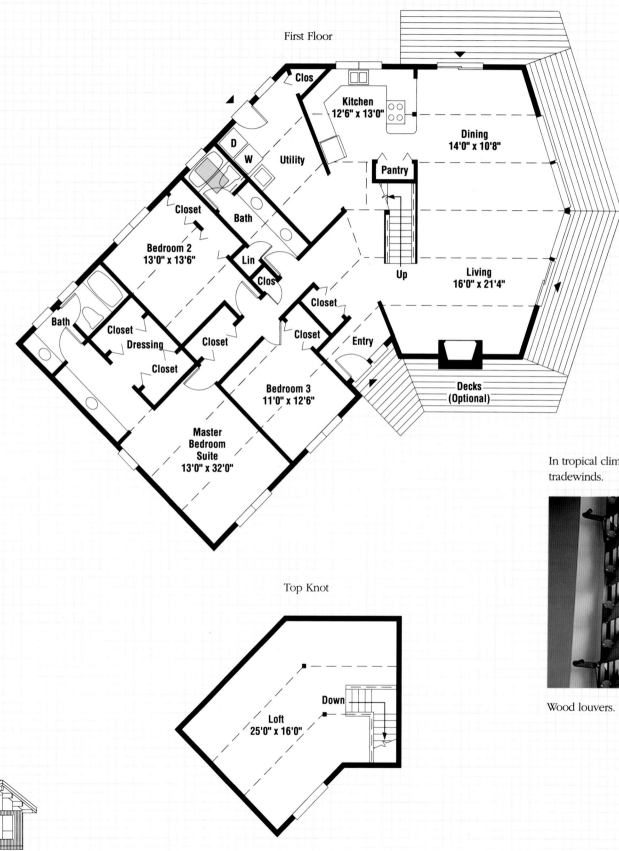

First Floor

Clos

Kitchen
12'6" x 13'0"

Dining
14'0" x 10'8"

D

W

Utility

Pantry

Bath

Living
16'0" x 21'4"

Up

Bedroom 2
13'0" x 13'6"

Lin

Clos

Closet

Bath

Closet

Dressing

Closet

Closet

Closet

Entry

Closet

Bedroom 3
11'0" x 12'6"

Master
Bedroom
Suite
13'0" x 32'0"

Decks
(Optional)

Top Knot

Down

Loft
25'0" x 16'0"

Solid cedar timbers and A.I.T.C. certified glu-laminated beams make a cozy living room. A wood framed sliding glass door opens to the deck.

In tropical climates, jalousie windows are popular to catch the tradewinds.

Wood louvers.   Glass louvers.

HERITAGE SERIES

HANCOCK

- 3 Bedrooms
- 2 Bathrooms
- Master bedroom on second floor
- 1577 sq. ft.
- First Floor: 981 sq. ft.
- Other Floor: 596 sq. ft.
- Overall Size: 37' x 30'

Photos of Heritages are
shown on pages 88-89.

▨ Shading indicates skylight or sunroom.

The Hancock has a 4-module gambrel (whereas this modification has a 5-module gambrel) with a wing on the left and a rear entry.

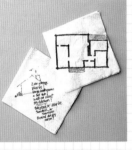

Start your dream home with a wish list and a rough layout. Instead of paper napkins, use the grid paper opposite page 152.

First Floor

Family
16'0" x 21'4"

Bath

Closet

Bedroom 3
12'0" x 10'6"

Linen

D
W

Closet

Kitchen
10'8" x 10'0"

Up

Dining
10'8" x 10'0"

Living
10'8" x 16'0"

Deck
(Optional)

Second Floor

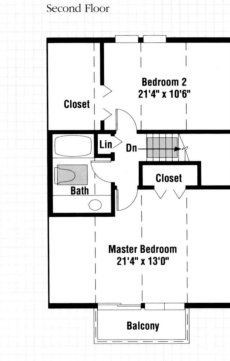

Closet

Bedroom 2
21'4" x 10'6"

Lin

Dn

Closet

Bath

Master Bedroom
21'4" x 13'0"

Balcony

Casement windows open for ventilation. Hardware, screen, flashing and trim are provided with all opening windows.

HERITAGE SERIES

YORKTOWN

- 3+ Bedrooms
- 2 3/4 Bathrooms
- Master bedroom on second floor
- 2148 sq. ft.
- First Floor: 1333 sq. ft.
- Other Floor: 815 sq. ft.
- Overall Size: 47' x 34'

Photos of Heritages are
shown on pages 88-89.

Shading indicates skylight or sunroom.

In this closeup, the mas-
sive Polar Cap 3 gambrel
roof stands out.

First Floor

Study
12'0" x 12'0"

Dining
14'8" x 12'0"

Kitchen
14'3" x 10'8"

Utility

W

D

2-Car Garage
22'0" x 21'4"
(Optional)

Closet

Up

Bath

Family
20'0" x 10'8"

Living
21'4" x 15'4"

Deck
(Optional)

Closet

Second Floor

Bedroom 2
12'8" x 13'0"

Bedroom 3
14'0" x 12'0"

Closet

Closet

Down

Bath

Lin

Closet

Master
Bedroom
Suite
21'4" x 18'0"

Open
to
Below

Balcony

The Yorktown has a 5-module, 2-
story gambrel with a side entry, and
a 1-story wing plus an optional 2-car
garage.

Ceilings in rooms on the
main floor of 2-story homes
feature A.I.T.C. certified glu-
laminated beams and wood
ceiling planks.

COLONIAL SERIES

MAYFLOWER

- 3 Bedrooms
- 2 Bathrooms
- Master bedroom on second floor
- 1875 sq. ft.
- First Floor: 1363 sq. ft.
- Other Floor: 512 sq. ft.
- Overall Size: 48' x 35'

Photos of Colonials are
shown on pages 90-91.

The owners built their Mayflower
on a walkout daylight basement and
added a screened-in porch.

First Floor

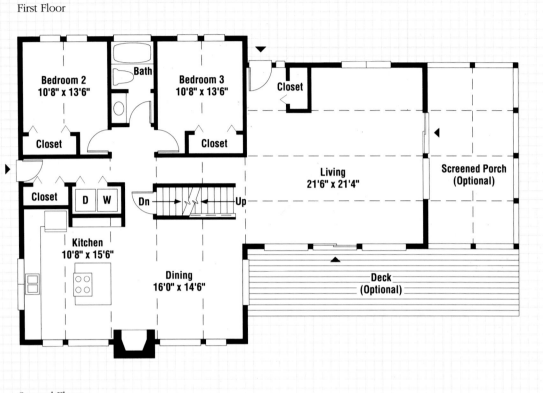

Bedroom 2
10'8" x 13'6"

Bath

Bedroom 3
10'8" x 13'6"

Closet

Closet

Closet

Closet

D   W

Dn   Up

Kitchen
10'8" x 15'6"

Living
21'6" x 21'4"

Screened Porch
(Optional)

Dining
16'0" x 14'6"

Deck
(Optional)

Second Floor

Master Bedroom
Suite
26'8" x 17'0"

Closet

Closet

Balcony   Down

Cathedral
Ceiling

Don't let insects keep you indoors
in summer. Build a covered porch
with tongue-and-groove cedar roof
liner and decking or floor covering of
your choice.

Here are two of Lindal's most popular staircase options:

Wood staircase/closed risers.
Comes with cedar 2x2 rail
for balusters and top rail.

Wood staircase/open risers.
Stringers are pre-routed to
accept the treads. Glu-laminated stringers and treads
are also available.

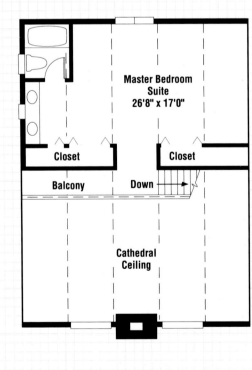

COLONIAL SERIES

PILGRIM

- 3 Bedrooms
- 2 1/2 Bathrooms
- Master bedroom on first floor
- 2313 sq. ft.
- First Floor: 1330 sq. ft.
- Other Floor: 983 sq. ft.
- Overall Size: 37' x 48'

Photos of Colonials are shown on pages 90-91.

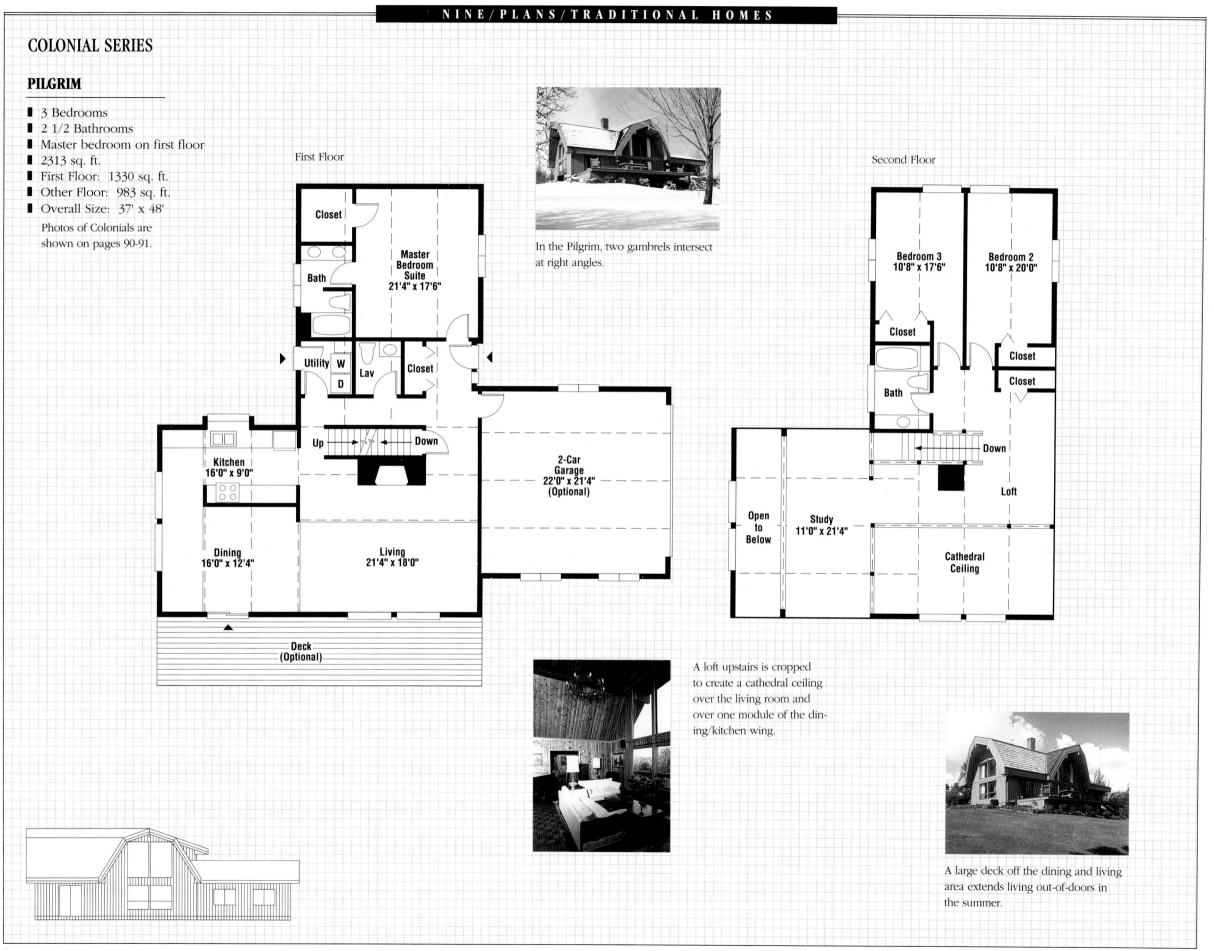

First Floor

Closet

Master Bedroom Suite
21'4" x 17'6"

Bath

Utility | **W D** | **Lav** | **Closet**

Up | **Down**

Kitchen
16'0" x 9'0"

2-Car Garage
22'0" x 21'4"
(Optional)

Dining
16'0" x 12'4"

Living
21'4" x 18'0"

Deck (Optional)

In the Pilgrim, two gambrels intersect at right angles.

Second Floor

Bedroom 3
10'8" x 17'6"

Bedroom 2
10'8" x 20'0"

Closet

Closet

Closet

Bath

Down

Loft

Open to Below

Study
11'0" x 21'4"

Cathedral Ceiling

A loft upstairs is cropped to create a cathedral ceiling over the living room and over one module of the dining/kitchen wing.

A large deck off the dining and living area extends living out-of-doors in the summer.

LIBERTY SERIES

INDEPENDENCE

- 3 Bedrooms
- 2 Bathrooms
- Master bedroom on first floor
- 1608 sq. ft.
- First Floor: 938 sq. ft.
- Other Floor: 670 sq. ft.
- Overall Size: 42' x 23'

Photos of Liberties are shown on pages 92-93.

░░ Shading indicates skylight or sunroom.

In the Liberty series a 2-story picture window dormer breaks up the ridge side of the gambrel roof, flooding the interior with light. Note: Stairs in photo are in different location than in plan.

First Floor

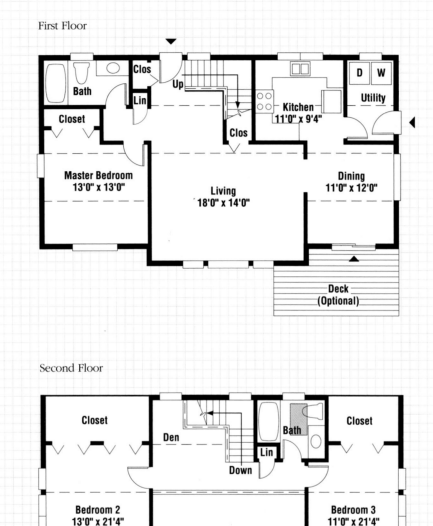

Master Bedroom 13'0" x 13'0"

Living 18'0" x 14'0"

Kitchen 11'0" x 9'4"

Dining 11'0" x 12'0"

Bath
Clos
Lin
Up
Clos
Closet
D W
Utility

Deck (Optional)

Second Floor

Closet

Den

Bath

Closet

Down

Lin

Bedroom 2 13'0" x 21'4"

Bedroom 3 11'0" x 21'4"

Cathedral Ceiling

A wood framed 3-dimensional garden window is a delightful option for over the kitchen sink.

This detail shows the venting side panel (with screen) on a garden window, which opens to catch a summer breeze.

LIBERTY SERIES

CHESAPEAKE

- 3 Bedrooms
- 3 Bathrooms
- Master bedroom on second floor
- 2075 sq. ft.
- First Floor: 1128 sq. ft.
- Other Floor: 947 sq. ft.
- Overall Size: 45' x 27'

Photos of Liberties are shown on pages 92-93.

The 2-story picture window dormer is the outstanding feature of the Liberty series. Note: Stairs in photo are in different location than in plan.

First Floor

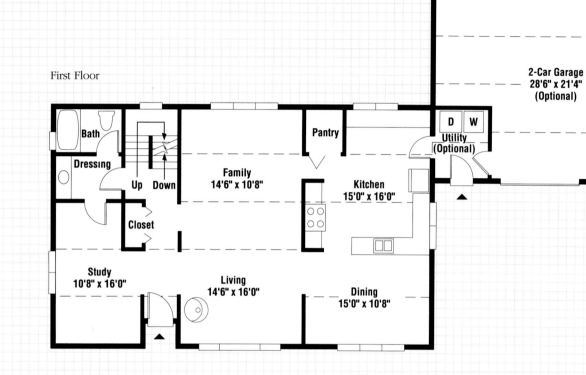

Bath

Dressing

Up Down

Closet

Study
10'8" x 16'0"

Family
14'6" x 10'8"

Living
14'6" x 16'0"

Pantry

Kitchen
15'0" x 16'0"

Dining
15'0" x 10'8"

D W

Utility
(Optional)

2-Car Garage
28'6" x 21'4"
(Optional)

The Chesapeake offers a recessed front entry, as opposed to the Independence, which has a rear entry.

Second Floor

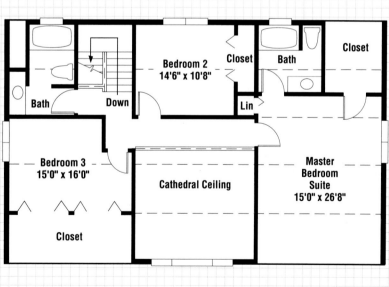

Bath

Down

Bedroom 3
15'0" x 16'0"

Closet

Cathedral Ceiling

Bedroom 2
14'6" x 10'8"

Closet

Lin

Bath

Closet

Master
Bedroom
Suite
15'0" x 26'8"

Think about awning windows; they open from the bottom – permitting ventilation, even when it's raining outside.

FARMHOUSE SERIES

PIONEER

- 3 Bedrooms
- 2 1/2 Bathrooms
- Master bedroom on first floor
- 1860 sq. ft.
- First Floor: 1067 sq. ft.
- Second Floor: 793 sq. ft.
- Overall Size: 40' x 27'

Photos of Farmhouses are shown on pages 94-95.

▨ Shading indicates skylight or sunroom.

Rustic Maple hardwood flooring harmonizes with the farmhouse motif.

First Floor

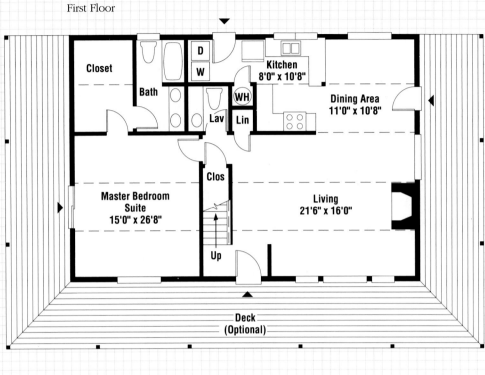

Closet

Bath

Kitchen
8'0" x 10'8"

D

W

WH

Lav | Lin

Dining Area
11'0" x 10'8"

Clos

Master Bedroom
Suite
15'0" x 26'8"

Living
21'6" x 16'0"

Up

Deck
(Optional)

The Pioneer has a wrap-around covered deck on three sides, and a central entry way.

Second Floor

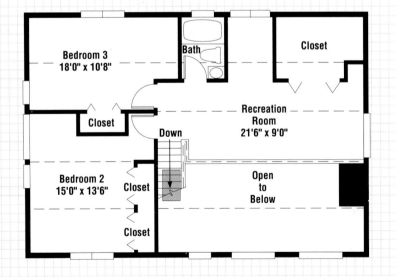

Bedroom 3
18'0" x 10'8"

Bath

Closet

Closet

Recreation
Room
21'6" x 9'0"

Down

Bedroom 2
15'0" x 13'6"

Closet

Open
to
Below

Closet

A covered deck is ideal for extending living out-of-doors and enjoying the view.

Important:
Farmhouse models are shown with Round Log siding. Please note that all plans can be ordered in Lindal, Clapboard, Round Log or Justus specifications.

FARMHOUSE SERIES

HOMESTEAD

- 3 Bedrooms
- 2 Bathrooms
- Master bedroom on first floor
- 1719 sq. ft.
- First Floor: 976 sq. ft.
- Second Floor: 743 sq. ft.
- Overall Size: 39' x 30'

Photos of Farmhouses are shown on pages 94-95.

The Homestead, unlike the other two plans, is not rectangular; it is 'L' shaped. This is a detail of the optional log look corners.

First Floor

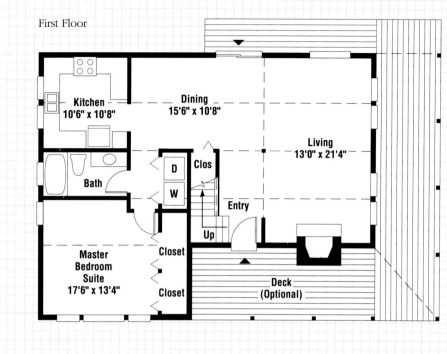

Kitchen 10'6" x 10'8"

Dining 15'6" x 10'8"

Living 13'0" x 21'4"

Bath

D

W

Clos

Entry

Up

Master Bedroom Suite 17'6" x 13'4"

Closet

Closet

Deck (Optional)

Second Floor

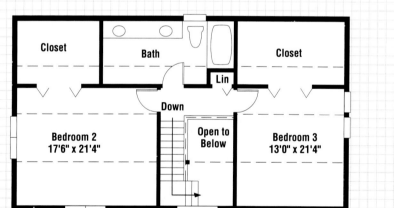

Closet

Bath

Closet

Lin

Down

Bedroom 2 17'6" x 21'4"

Open to Below

Bedroom 3 13'0" x 21'4"

French doors can be added to any plan. These photos show them both with and without removable grids.

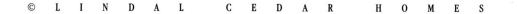

FARMHOUSE SERIES

HARVESTER

- 4 Bedrooms
- 2 1/2 Bathrooms
- Master bedroom on first floor
- 2368 sq. ft.
- First Floor: 1344 sq. ft.
- Other Floor: 1024 sq. ft.
- Overall Size: 42' x 32'

Photos of Farmhouses are shown on pages 94-95.

A covered deck wraps around all four sides of the Harvester. Here's another detail of the optional log look corners.

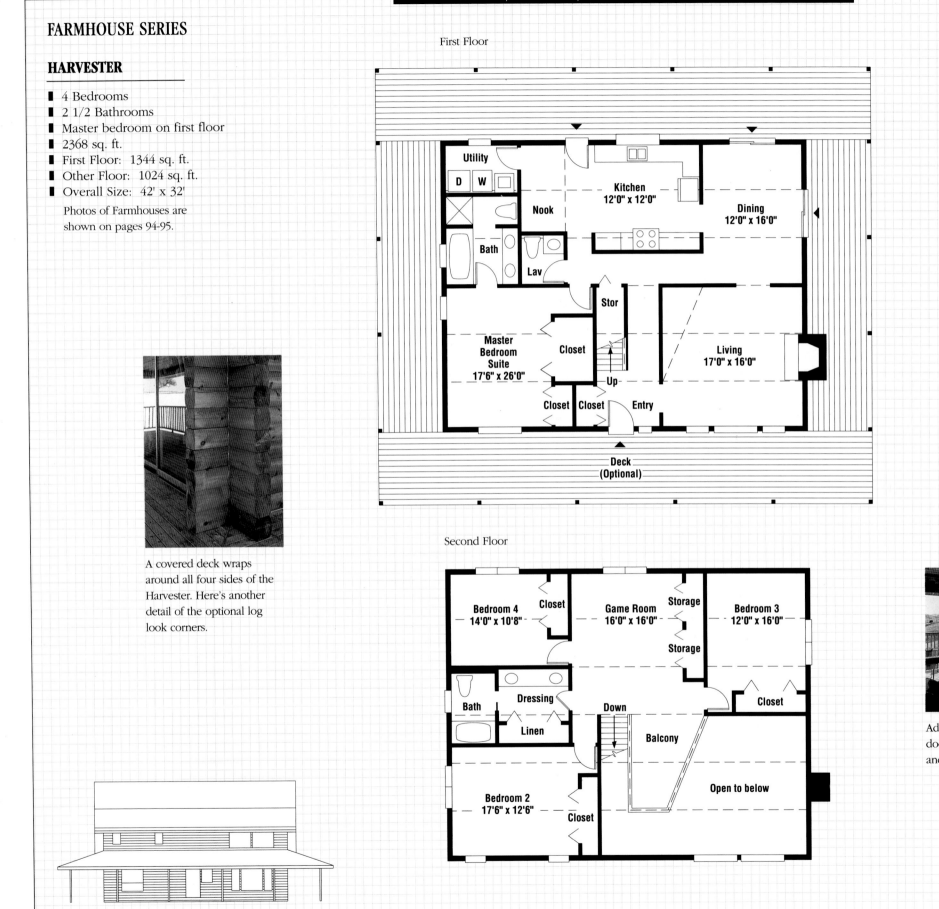

First Floor

Utility

D W

Kitchen
12'0" x 12'0"

Nook

Dining
12'0" x 16'0"

Bath

Lav

Stor

Master Bedroom Suite
17'6" x 26'0"

Closet

Living
17'0" x 16'0"

Up

Closet Closet Entry

Deck (Optional)

Second Floor

Bedroom 4
14'0" x 10'8"

Closet

Game Room
16'0" x 16'0"

Storage

Bedroom 3
12'0" x 16'0"

Storage

Bath

Dressing

Down

Closet

Linen

Balcony

Bedroom 2
17'6" x 12'6"

Open to below

Closet

Consider a skywall either for your kitchen or bathroom. Skywalls are designed to sit over a 40" high base counter and they can be inset into the roof to a depth of 3 or 6 ft.

Add as many windows and opening doors as you wish to enjoy your view and deck living.

NEW ENGLAND SERIES

PATRIOT

- 3 Bedrooms
- 2 Bathrooms
- Master bedroom on first floor
- 2029 sq. ft.
- First Floor: 1075 sq. ft.
- Other Floor: 954 sq. ft.
- Overall Size: 40' x 29'

Photos of New Englands are shown on pages 96-97.

Double hung windows are a popular choice for traditional homes. Both halves move. Removable grids are available.

First Floor

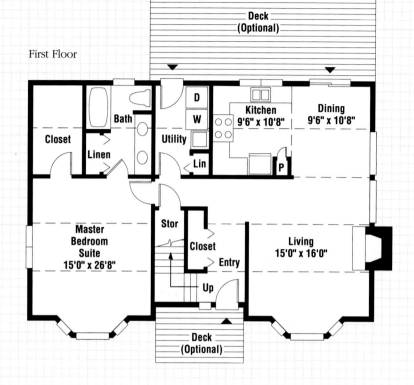

Here is a fixed window over an awning window – a functional arrangement in bathrooms.

Second Floor

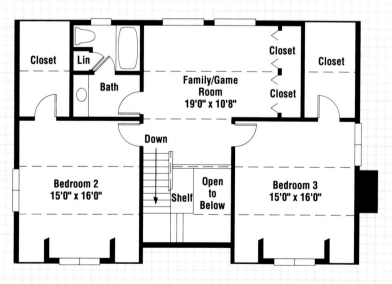

Eyebrow dormers suit the traditional look of the New England series.

Important:
New England models are shown with Clapboard siding. Please note that all plans are available in Lindal, Clapboard, Round Log or Justus specifications.

NEW ENGLAND SERIES

WHALER

- 4 Bedrooms
- 2 1/2 Bathrooms
- Master bedroom on second floor
- 2293 sq. ft.
- First Floor: 1152 sq. ft.
- Other Floor: 1141 sq. ft.
- Overall Size: 37' x 32'

Photos of New Englands are shown on pages 96-97.

Shading indicates skylight or sunroom.

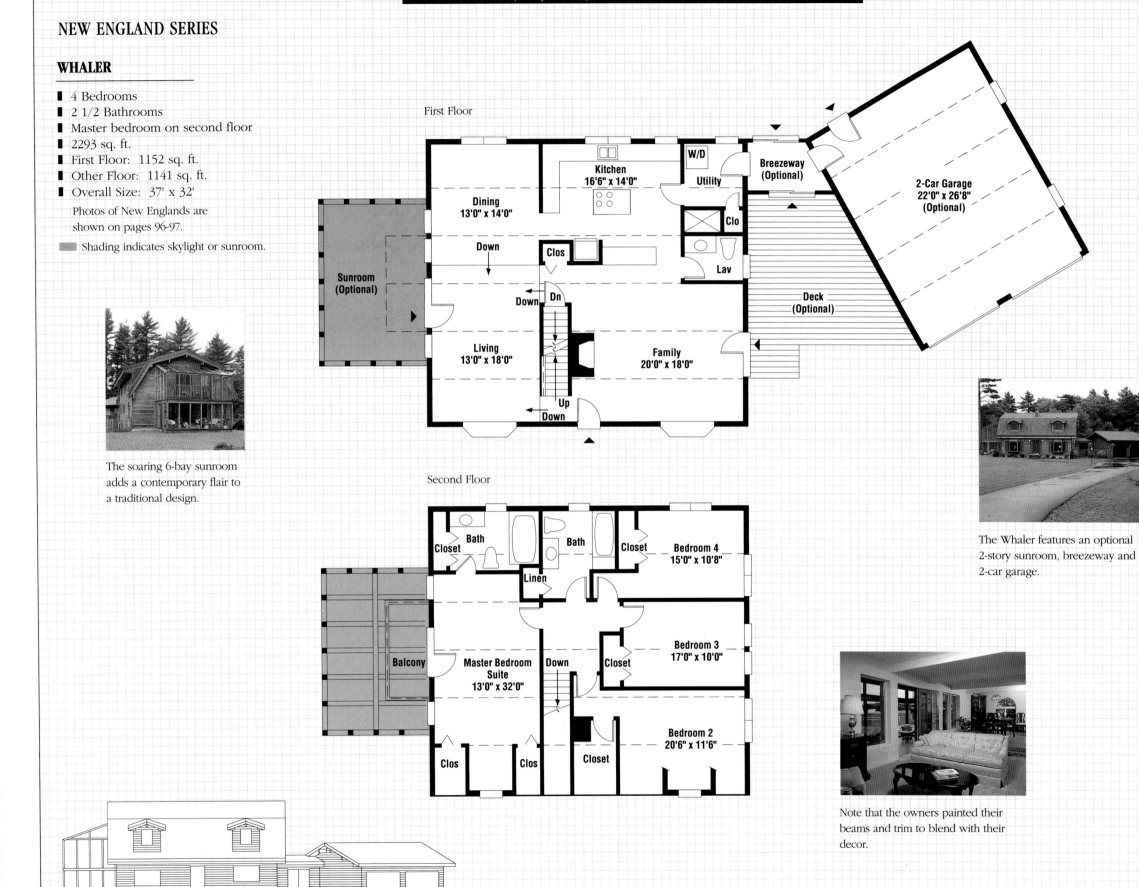

First Floor

Dining
13'0" x 14'0"

Kitchen
16'6" x 14'0"

W/D

Utility

Breezeway
(Optional)

2-Car Garage
22'0" x 26'8"
(Optional)

Down

Clos

Clo

Lav

Sunroom
(Optional)

Down

Dn

Down

Living
13'0" x 18'0"

Family
20'0" x 18'0"

Deck
(Optional)

Up
Down

The soaring 6-bay sunroom adds a contemporary flair to a traditional design.

The Whaler features an optional 2-story sunroom, breezeway and 2-car garage.

Second Floor

Closet

Bath

Bath

Closet

Bedroom 4
15'0" x 10'8"

Linen

Balcony

Master Bedroom
Suite
13'0" x 32'0"

Down

Closet

Bedroom 3
17'0" x 10'0"

Clos

Clos

Closet

Bedroom 2
20'6" x 11'6"

Note that the owners painted their beams and trim to blend with their decor.

NEW ENGLAND SERIES

CONSTITUTION

- 3+ Bedrooms
- 3 Bathrooms
- Master bedroom on second floor
- 3054 sq. ft.
- First Floor: 1551 sq. ft.
- Other Floor: 1503 sq. ft.
- Overall Size: 44' x 39'

Photos of New Englands are shown on pages 96-97.

A Polar Cap 3 roof with R-63 insulation is the ultimate in energy efficiency. Hint: You can spot Polar 3 roofs by the 3 layers of fascia.

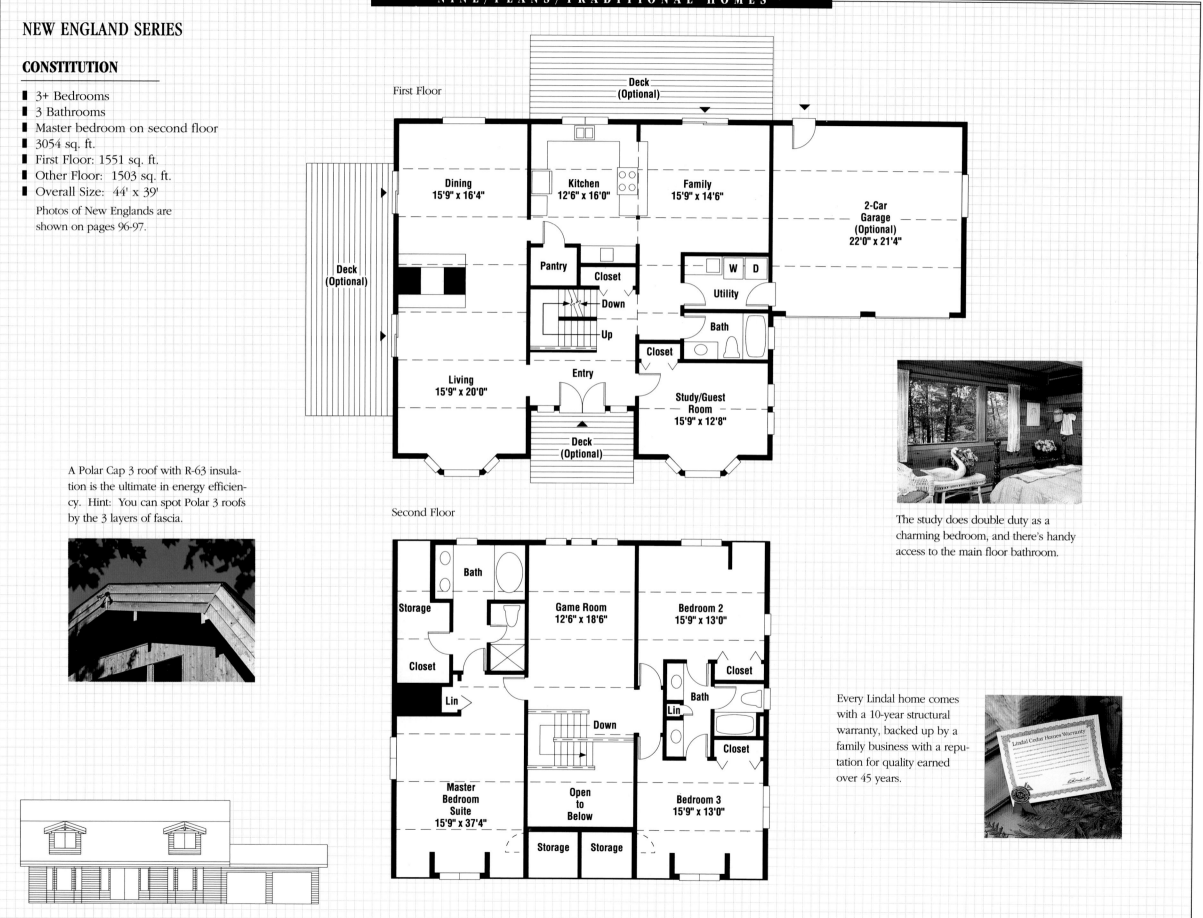

First Floor

Deck (Optional)

Deck (Optional)

Dining
15'9" x 16'4"

Kitchen
12'6" x 16'0"

Family
15'9" x 14'6"

2-Car Garage (Optional)
22'0" x 21'4"

Pantry

Closet

W D

Utility

Down

Bath

Up

Closet

Living
15'9" x 20'0"

Entry

Study/Guest Room
15'9" x 12'8"

Deck (Optional)

Second Floor

Bath

Storage

Game Room
12'6" x 18'6"

Bedroom 2
15'9" x 13'0"

Closet

Closet

Lin

Bath

Lin

Closet

Down

Master Bedroom Suite
15'9" x 37'4"

Open to Below

Bedroom 3
15'9" x 13'0"

Storage

Storage

The study does double duty as a charming bedroom, and there's handy access to the main floor bathroom.

Every Lindal home comes with a 10-year structural warranty, backed up by a family business with a reputation for quality earned over 45 years.

PROW SERIES

VENICE

- 2 Bedrooms
- 2 Bathrooms
- Master bedroom on second floor
- 1016 sq. ft.
- First Floor: 688 sq. ft.
- Other Floor 328 sq. ft.
- Overall Size: 21' x 34'

Photos of Prows are shown on page 100.

Shading indicates skylight or sunroom.

VIKING

- 3 Bedrooms
- 2 Bathrooms
- Master bedroom on second floor
- 1876 sq. ft.
- First Floor: 1090 sq. ft.
- Other Floor: 786 sq. ft.
- Overall Size: 29' x 42'

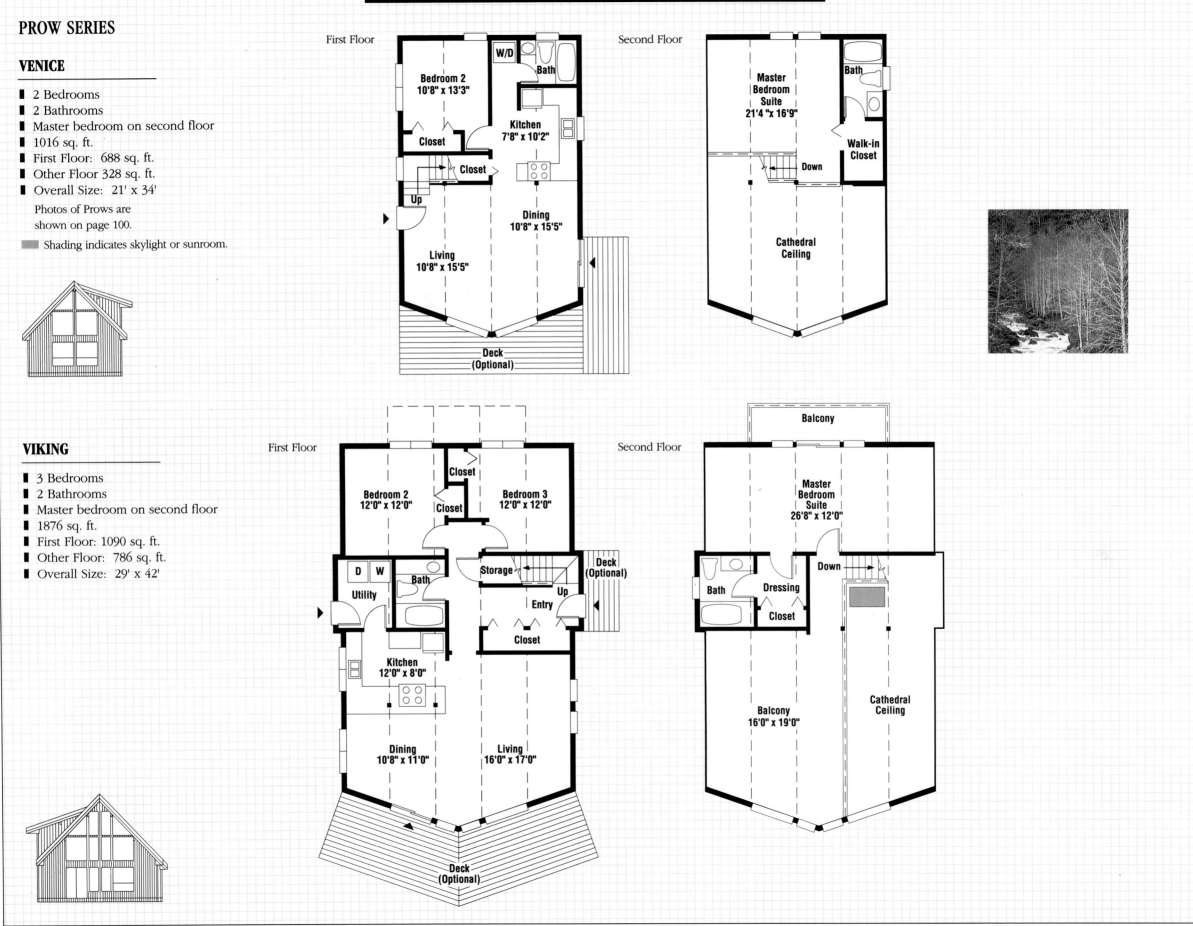

First Floor

Bedroom 2
10'8" x 13'3"

W/D

Bath

Closet

Kitchen
7'8" x 10'2"

Closet

Up

Dining
10'8" x 15'5"

Living
10'8" x 15'5"

Deck
(Optional)

Second Floor

Bath

Master Bedroom Suite
21'4" x 16'9"

Walk-in Closet

Down

Cathedral Ceiling

First Floor

Closet

Bedroom 2
12'0" x 12'0"

Closet

Bedroom 3
12'0" x 12'0"

D   W

Bath

Utility

Storage

Deck (Optional)

Up

Entry

Closet

Kitchen
12'0" x 8'0"

Dining
10'8" x 11'0"

Living
16'0" x 17'0"

Deck
(Optional)

Second Floor

Balcony

Master Bedroom Suite
26'8" x 12'0"

Down

Bath

Dressing

Closet

Balcony
16'0" x 19'0"

Cathedral Ceiling

PROW SERIES

VISTA

- 2 Bedrooms
- 1 Bathroom
- Master bedroom on first floor
- 1246 sq. ft.
- First Floor: 790 sq. ft.
- Other Floor: 456 sq. ft.
- Overall Size: 21' x 39'

Photos of Prows are shown on page 100.

Shading indicates skylight or sunroom.

VOLARE

- 3 Bedrooms
- 2 Bathrooms
- Master bedroom on second floor
- 1624 sq. ft.
- First Floor: 1019 sq. ft.
- Other Floor: 605 sq. ft.
- Overall Size: 27' x 40'

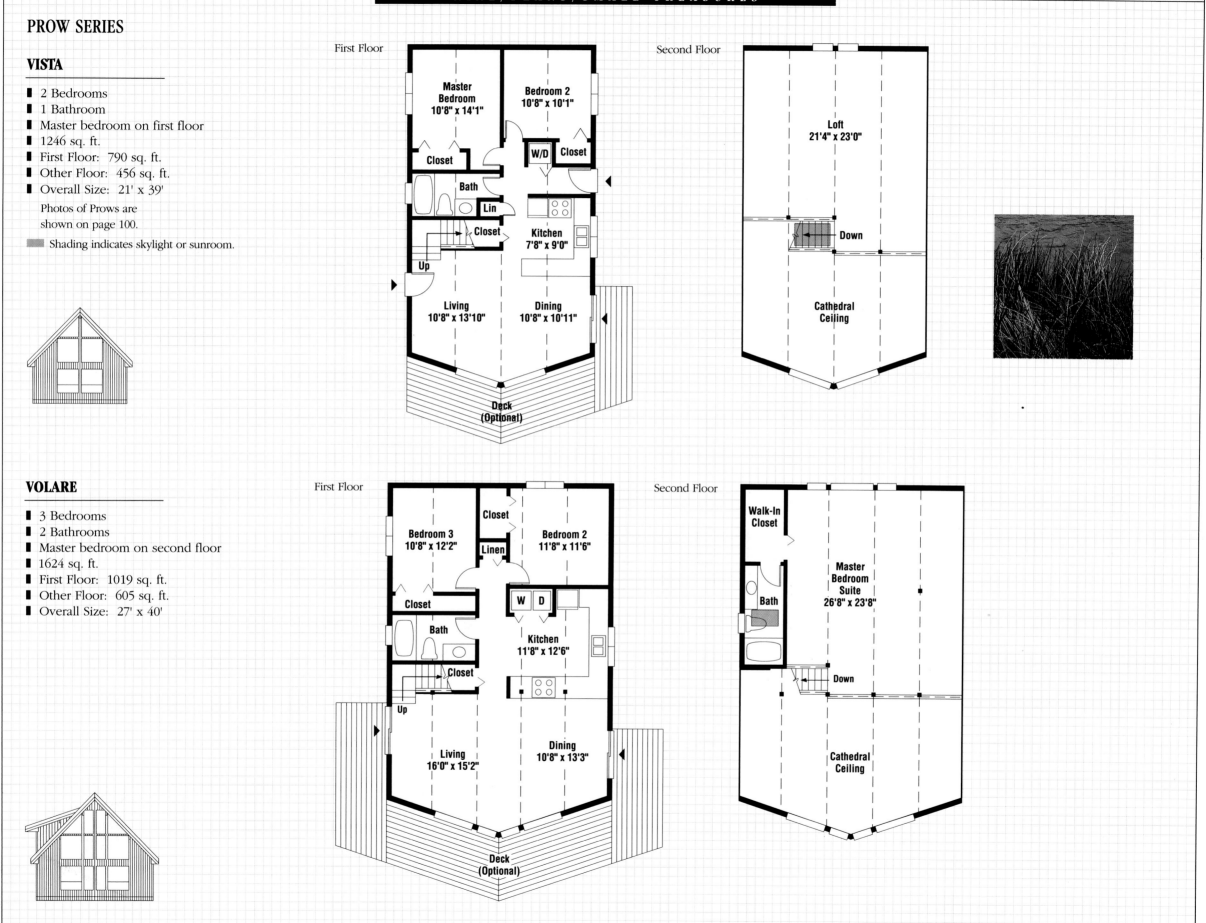

First Floor

Master Bedroom 10'8" x 14'1"

Bedroom 2 10'8" x 10'1"

Closet

W/D   Closet

Bath

Lin

Up   Closet

Kitchen 7'8" x 9'0"

Living 10'8" x 13'10"

Dining 10'8" x 10'11"

Deck (Optional)

Second Floor

Loft 21'4" x 23'0"

Down

Cathedral Ceiling

First Floor

Closet

Bedroom 3 10'8" x 12'2"

Linen

Bedroom 2 11'8" x 11'6"

Closet

Bath

W   D

Closet

Up

Kitchen 11'8" x 12'6"

Living 16'0" x 15'2"

Dining 10'8" x 13'3"

Deck (Optional)

Second Floor

Walk-In Closet

Bath

Master Bedroom Suite 26'8" x 23'8"

Down

Cathedral Ceiling

PROW SERIES

VEGAS

- 1 Bedroom
- 1 Bathroom
- Master bedroom on first floor
- 766 sq. ft.
- First Floor: 507 sq. ft.
- Other Floor: 259 sq. ft.
- Overall Size: 21' x 26'

Photos of Prows are shown on page 100.

VICTORY

- 3 Bedrooms
- 2 1/2 Bathrooms
- Master bedroom on first floor
- 2300 sq. ft.
- First Floor: 1461 sq. ft.
- Other Floor: 839 sq. ft.
- Overall Size: 32' x 48'

First Floor

Master Bedroom 12'6" x 10'2"
Closet
Bath
Kitchen 5'6" x 7'9"
Storage
Up
Dining 6'6" x 10'2"
Living 14'10" x 11'2"
Deck (Optional)

Second Floor

Loft 21'4" x 13'6"
Down
Cathedral Ceiling

First Floor

Bath
WH
Linen
Lav
D W Clo
Entry
Clos
Kitchen 16'0" x 9'0"
Closet
Master Bedroom Suite 32'0" x 16'0"
Clos
Up
Closet
Living 16'0" x 23'0"
Dining 16'0" x 14'0"
Deck (Optional)

Second Floor

Bedroom 2 14'0" x 18'6"
Bath
Bedroom 3 13'0" x 16'0"
Closet
Linen
Down
Closet
Rec Room 16'0" x 14'0"
Cathedral Ceiling

CHALET SERIES

MATTERHORN

- 2 Bedrooms
- 1 1/2 Bathrooms
- Master bedroom on second floor
- 1173 sq. ft.
- First Floor: 608 sq. ft.
- Other Floor: 565 sq. ft.
- Overall Size: 21' x 29'

Photos of Chalets are
shown on page 101.

Shading indicates skylight or sunroom.

GRENOBLE

- 2 Bedrooms
- 1 Bathroom
- Master bedroom on first floor
- 1187 sq. ft.
- First Floor: 806 sq. ft.
- Other Floor: 381 sq. ft.
- Overall Size: 28' x 30'

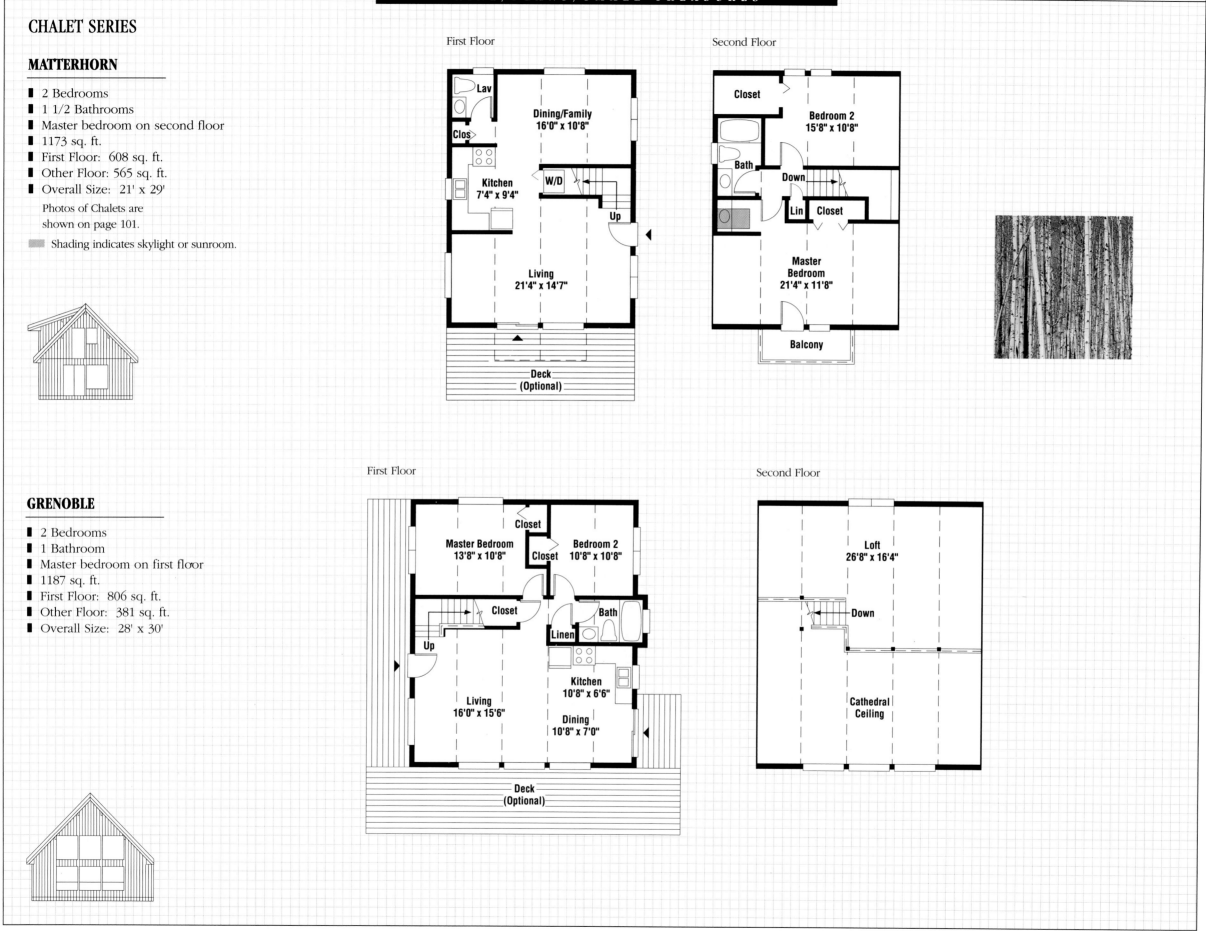

First Floor

Lav

Dining/Family
16'0" x 10'8"

Clos

Kitchen
7'4" x 9'4"

W/D

Up

Living
21'4" x 14'7"

Deck
(Optional)

Second Floor

Closet

Bedroom 2
15'8" x 10'8"

Bath

Down

Lin   Closet

Master
Bedroom
21'4" x 11'8"

Balcony

First Floor

Closet

Master Bedroom
13'8" x 10'8"

Bedroom 2
10'8" x 10'8"

Closet

Closet

Up

Bath

Linen

Living
16'0" x 15'6"

Kitchen
10'8" x 6'6"

Dining
10'8" x 7'0"

Deck
(Optional)

Second Floor

Loft
26'8" x 16'4"

Down

Cathedral
Ceiling

CHALET SERIES

STOWE

- 2 Bedrooms
- 2 Bathrooms
- Master bedroom on second floor
- 1277 sq. ft.
- First Floor: 832 sq. ft.
- Other Floor: 445 sq. ft.
- Overall Size: 25' x 36'

Photos of Chalets are shown on page 101.

▨ Shading indicates skylight or sunroom.

TAHOE

- 3 Bedrooms
- 2 Bathrooms
- Master bedroom on second floor
- 1600 sq. ft.
- First Floor: 1039 sq. ft.
- Other Floor: 561 sq. ft.
- Overall Size: 27' x 40'

First Floor

Bedroom 2
12'4" x 10'0"

Closet

Utility — W / D

Kitchen
11'6" x 10'0"

Closet

Bath

Storage — Up

Dining
12'8" x 16'0"

Living
12'8" x 16'0"

Second Floor

Closet

Bath

Master Bedroom Suite
21'4" x 20'0"

Down

Cathedral Ceiling

First Floor

Bedroom 2
13'8" x 12'3"

Closet

Closet

Bedroom 3
10'8" x 12'3"

Bath

Linen

Storage — W / D

Closet

Up

Kitchen
10'8" x 8'3"

Living
16'0" x 17'7"

Dining
10'8" x 10'8"

Deck (Optional)

Second Floor

Master Bedroom Suite
26'8" x 18'9"

Closet

Bath

Down

Balcony

Cathedral Ceiling

SUMMIT SERIES

SHASTA

- 2 Bedrooms
- 2 Bathrooms
- Master bedroom on second floor
- 1488 sq. ft.
- First Floor: 1109 sq. ft.
- Other Floor: 379 sq. ft.
- Overall Size: 32' x 37'

Photos of Summits are
shown on pages 102-103.

▨ Shading indicates skylight or sunroom.

SIERRA

- 2 Bedrooms
- 1 Bathroom
- Master bedroom on first floor
- 1433 sq. ft.
- First Floor: 913 sq. ft.
- Other Floor 520 sq. ft.
- Overall Size: 27' x 37'

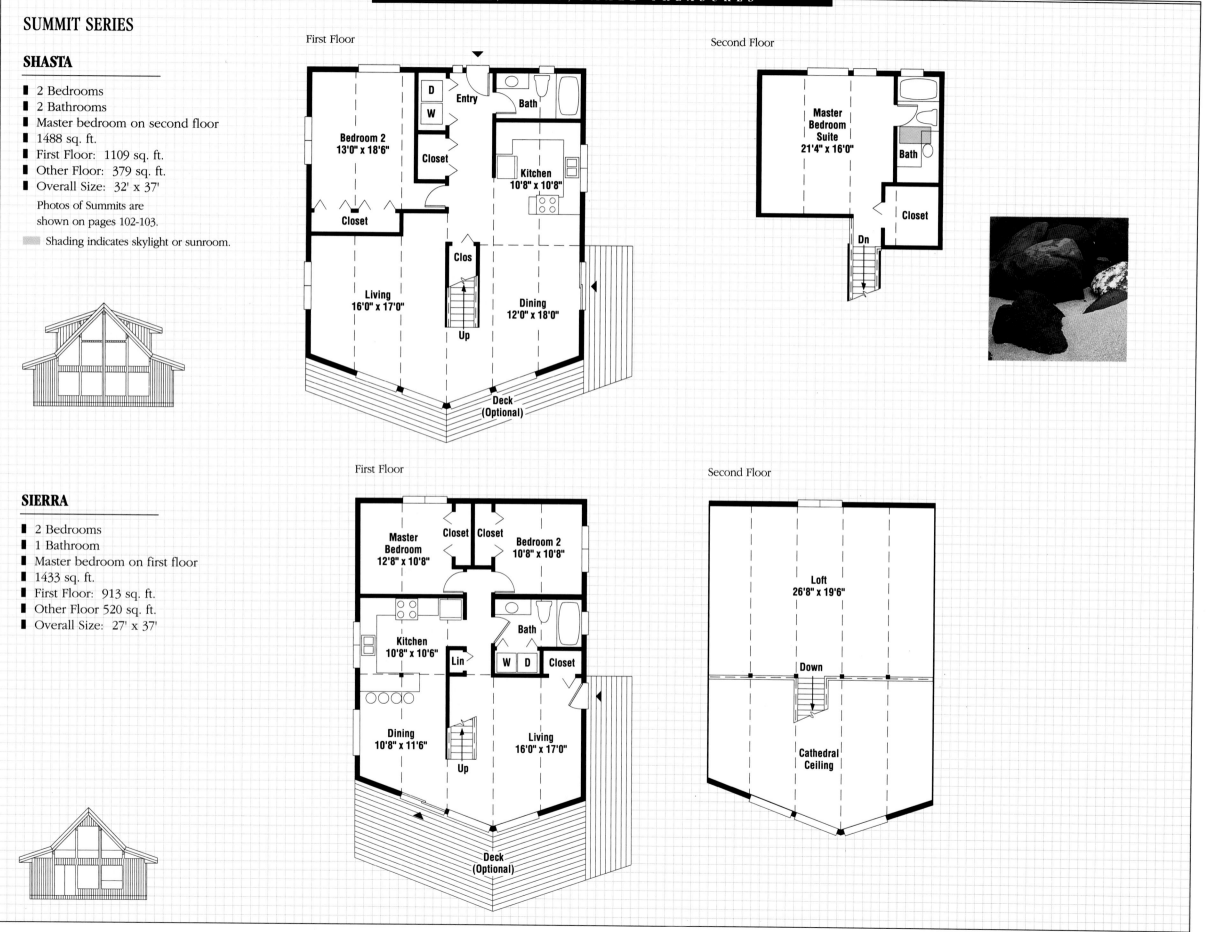

First Floor

D
W
Entry
Bath

Bedroom 2
13'0" x 18'6"

Closet

Kitchen
10'8" x 10'8"

Closet

Clos

Living
16'0" x 17'0"

Up

Dining
12'0" x 18'0"

Deck
(Optional)

Second Floor

Master
Bedroom
Suite
21'4" x 16'0"

Bath

Closet

Dn

First Floor

Master
Bedroom
12'8" x 10'8"

Closet   Closet

Bedroom 2
10'8" x 10'8"

Kitchen
10'8" x 10'6"

Bath

Lin

W   D   Closet

Dining
10'8" x 11'6"

Up

Living
16'0" x 17'0"

Deck
(Optional)

Second Floor

Loft
26'8" x 19'6"

Down

Cathedral
Ceiling

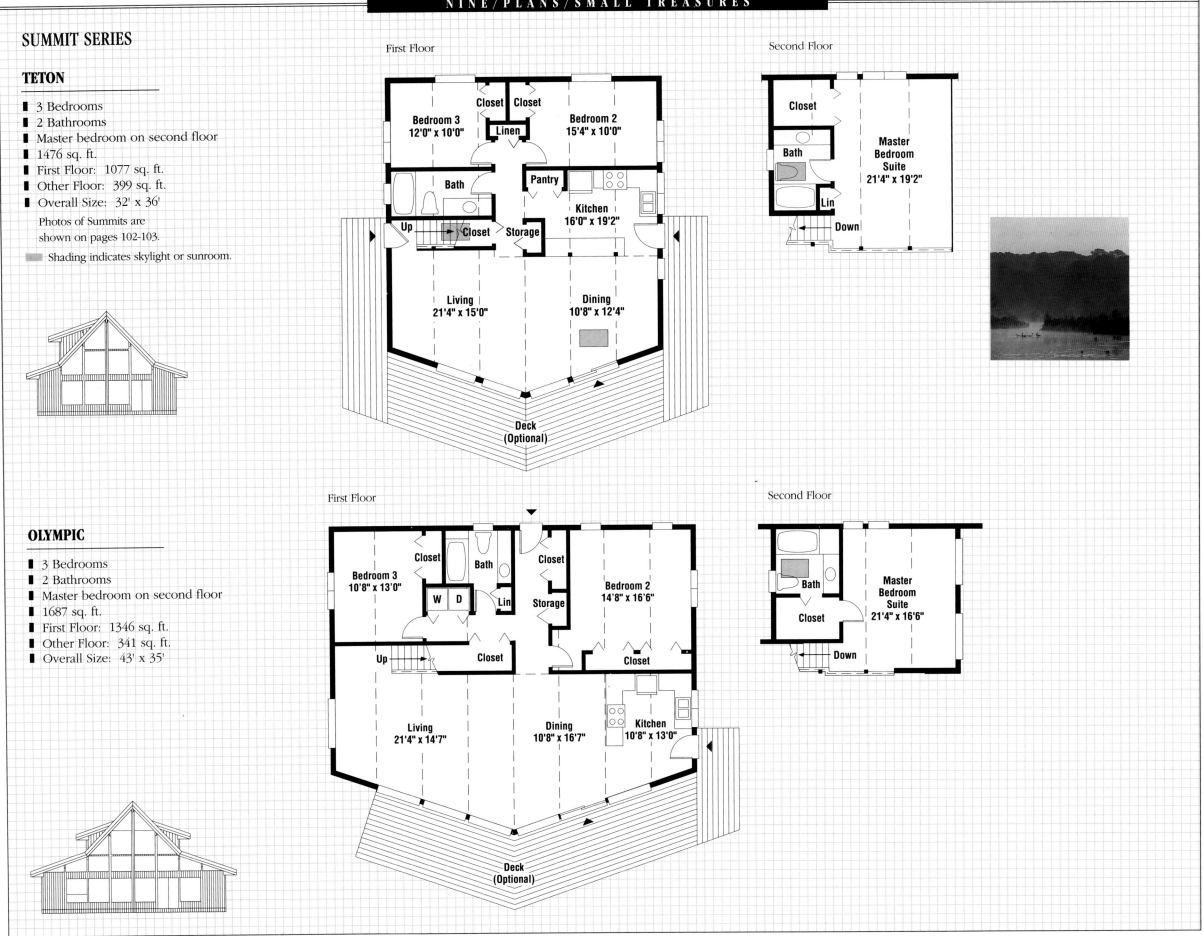

SUMMIT SERIES

TETON

- 3 Bedrooms
- 2 Bathrooms
- Master bedroom on second floor
- 1476 sq. ft.
- First Floor: 1077 sq. ft.
- Other Floor: 399 sq. ft.
- Overall Size: 32' x 36'

Photos of Summits are shown on pages 102-103.

▨ Shading indicates skylight or sunroom.

First Floor

Closet Closet

Bedroom 3
12'0" x 10'0"

Linen

Bedroom 2
15'4" x 10'0"

Bath

Pantry

Kitchen
16'0" x 19'2"

Up

Closet

Storage

Living
21'4" x 15'0"

Dining
10'8" x 12'4"

Deck
(Optional)

Second Floor

Closet

Bath

Lin

Master
Bedroom
Suite
21'4" x 19'2"

Down

OLYMPIC

- 3 Bedrooms
- 2 Bathrooms
- Master bedroom on second floor
- 1687 sq. ft.
- First Floor: 1346 sq. ft.
- Other Floor: 341 sq. ft.
- Overall Size: 43' x 35'

First Floor

Closet

Bath

Closet

Bedroom 3
10'8" x 13'0"

W D

Lin

Storage

Bedroom 2
14'8" x 16'6"

Up

Closet

Closet

Closet

Living
21'4" x 14'7"

Dining
10'8" x 16'7"

Kitchen
10'8" x 13'0"

Deck
(Optional)

Second Floor

Bath

Closet

Master
Bedroom
Suite
21'4" x 16'6"

Down

CONTEMPO PROW SERIES

CONSTELLATION

- 3 Bedrooms
- 2 Bathrooms
- Master bedroom on second floor
- 1326 sq. ft.
- First Floor: 1068 sq. ft.
- Other Floor: 258 sq. ft.
- Overall Size: 37' x 32'

Photos of Contempo Prows are shown on page 104.

Shading indicates skylight or sunroom.

First Floor

Bedroom 2
10'8" x 14'6"

Bath

Bedroom 3
10'8" x 14'6"

Kitchen
10'8" x 12'0"

Closet

Closet

Closet

Linen/Closet   Up

Living
21'4" x 13'6"

Dining
16'0" x 13'6"

Deck
(Optional)

Top Knot

Bath

Master
Bedroom
Suite
16'0" x 17'9"

Closet

Closet   Down

NEPTUNE

- 3 Bedrooms
- 2 Bathrooms
- Master bedroom on second floor
- 1887 sq. ft.
- First Floor: 1517 sq. ft.
- Other Floor: 370 sq. ft.
- Overall Size: 43' x 39'

First Floor

Kitchen
10'8" x 15'0"

Bedroom 3
10'8" x 15'0"

Closet

Bath

Bedroom 2
10'8" x 18'6"

D
W

Closet

Up   Storage

Closet

Dining
10'8" x 19'0"

Living
32'0" x 22'0"

Deck
(Optional)

Top Knot

Master
Bedroom
Suite
21'4" x 18'6"

Bath

Closet

Down

CONTEMPO PROW SERIES

POLARIS

- 2+ Bedrooms
- 3 Bathrooms
- Master bedroom on first floor
- 1832 sq. ft.
- First Floor: 1485 sq. ft.
- Other Floor: 347 sq. ft.
- Overall Size: 43' x 49'

Photos of Contempo Prows are shown on page 104.

▨ Shading indicates skylight or sunroom.

CORONA

- 2 Bedrooms
- 1 Bathroom
- 1173 sq. ft.
- Overall Size: 32' x 39'

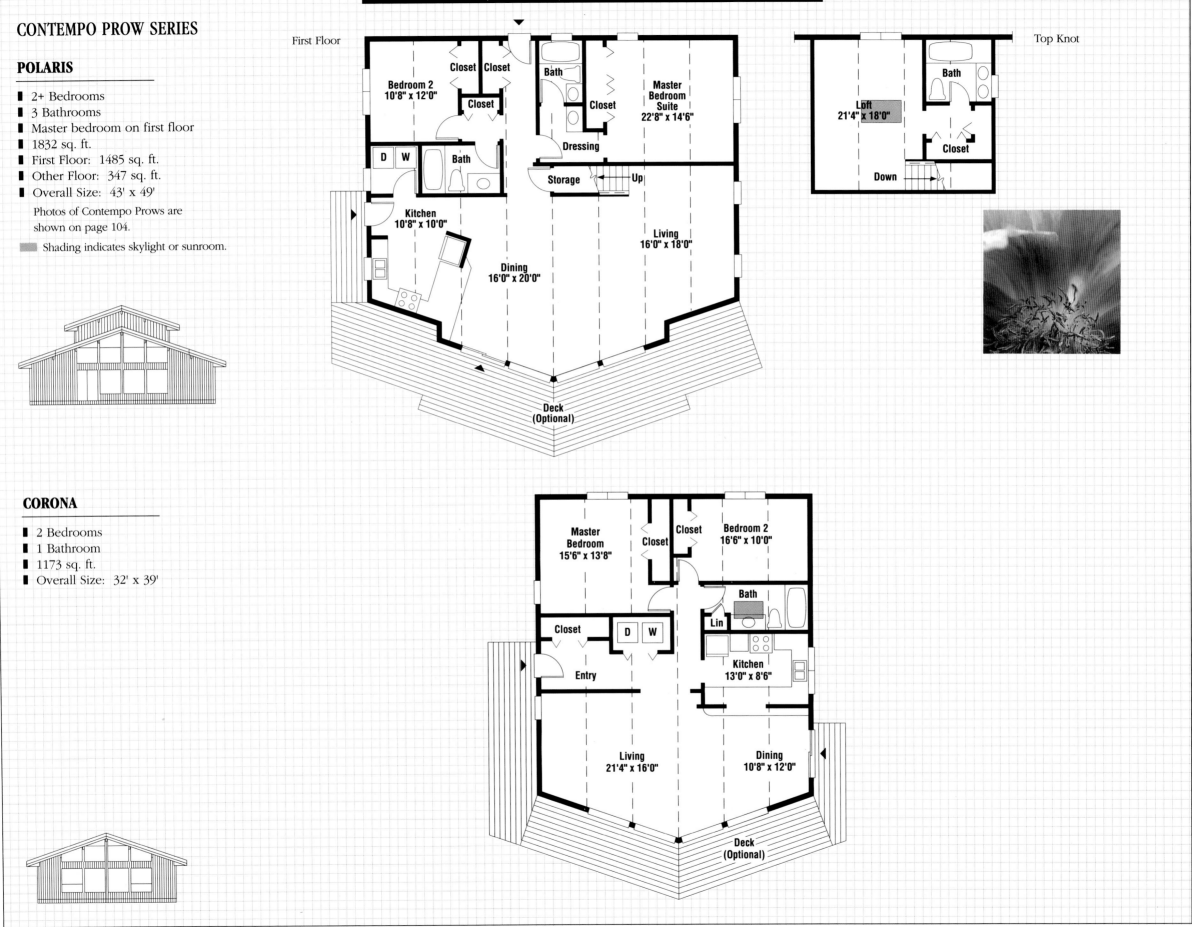

First Floor

Bedroom 2
10'8" x 12'0"

Closet Closet

Closet

Bath

Closet

Master
Bedroom
Suite
22'8" x 14'6"

D W Bath

Kitchen
10'8" x 10'0"

Dressing

Storage Up

Living
16'0" x 18'0"

Dining
16'0" x 20'0"

Deck
(Optional)

Top Knot

Bath

Loft
21'4" x 18'0"

Closet

Down

Master
Bedroom
15'6" x 13'8"

Closet

Closet

Bedroom 2
16'6" x 10'0"

Closet

Bath

Closet D W Lin

Kitchen
13'0" x 8'6"

Entry

Living
21'4" x 16'0"

Dining
10'8" x 12'0"

Deck
(Optional)

VIEW SERIES

CAPRI

- 2 Bedrooms
- 1 Bathroom
- 720 sq. ft.
- Overall Size: 27' x 27'

  Photos of Views are shown on page 105.

MALIBU

- 3 Bedrooms
- 2 Bathrooms
- 1016 sq. ft.
- Overall Size: 45' x 27'

This wing can be added to the Capri now, or later, to add more living space overall and put the master bedroom suite in its own wing.

DAYTONA

- 2 Bedrooms
- 1 Bathroom
- 907 sq. ft.
- Overall Size: 27' x 34'

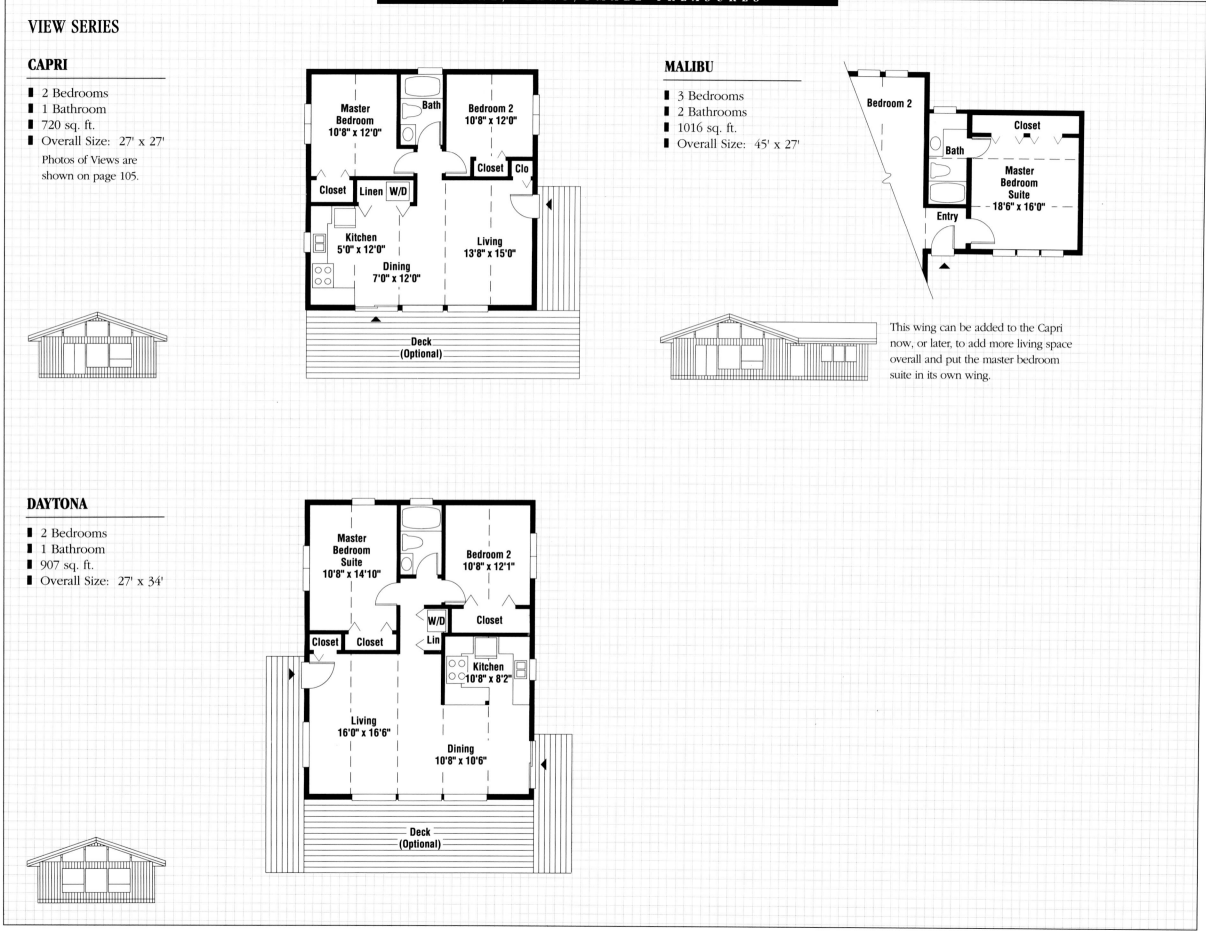

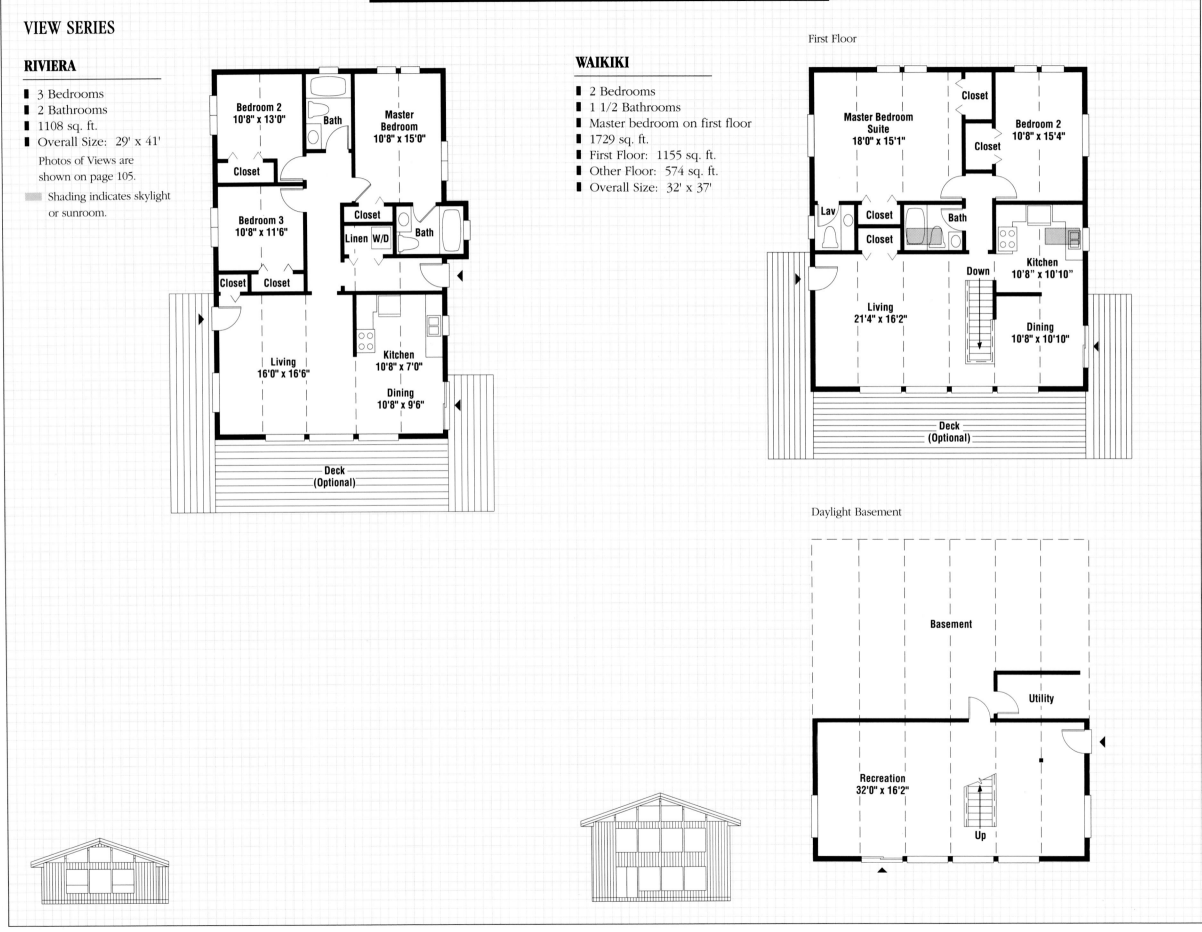

VIEW SERIES

RIVIERA

- 3 Bedrooms
- 2 Bathrooms
- 1108 sq. ft.
- Overall Size: 29' x 41'

  Photos of Views are shown on page 105.

  Shading indicates skylight or sunroom.

Bedroom 2
10'8" x 13'0"

Bath

Master Bedroom
10'8" x 15'0"

Closet

Bedroom 3
10'8" x 11'6"

Closet

Linen **W/D** **Bath**

Closet **Closet**

Living
16'0" x 16'6"

Kitchen
10'8" x 7'0"

Dining
10'8" x 9'6"

Deck
(Optional)

WAIKIKI

- 2 Bedrooms
- 1 1/2 Bathrooms
- Master bedroom on first floor
- 1729 sq. ft.
- First Floor: 1155 sq. ft.
- Other Floor: 574 sq. ft.
- Overall Size: 32' x 37'

First Floor

Master Bedroom Suite
18'0" x 15'1"

Closet

Bedroom 2
10'8" x 15'4"

Closet

Lav

Closet

Bath

Closet

Kitchen
10'8" x 10'10"

Living
21'4" x 16'2"

Down

Dining
10'8" x 10'10"

Deck
(Optional)

Daylight Basement

Basement

Utility

Recreation
32'0" x 16'2"

Up

PANORAMA SERIES

ACAPULCO

- 4 Bedrooms
- 2 Bathrooms
- Master bedroom on second floor
- 1184 sq. ft.
- First Floor: 952 sq. ft.
- Other Floor: 232 sq. ft.
- Overall Size: 37' x 26'

Photos of Panoramas are shown on pages 106-107.

▨ Shading indicates skylight or sunroom.

First Floor

Kitchen 10'8" x 12'2"
Closet
Bedroom 3 10'8" x 12'2"
Bath
Bedroom 4 10'6" x 9'10"
Linen
Closet
Closet
Up
Storage
Dining 10'8" x 12'10"
Living 16'0" x 9'10"
Bedroom 2 10'8" x 12'10"
Deck (Optional)

Top Knot

Master Bedroom Suite 16'0" x 15'5"
Bath
Down
Closet

MONACO

- 3 Bedrooms
- 1 3/4 Bathrooms
- Master bedroom on first floor
- 1555 sq. ft.
- First Floor: 1337 sq. ft.
- Other Floor: 218 sq. ft.
- Overall Size: 43' x 37'

First Floor

Bedroom 3 10'8" x 10'6"
Closet
Pan
Kitchen 16'0" x 8'6"
Dining 10'8" x 12'8"
Closet
Up
W/D
Closet
Bath
Linen
Living 21'4" x 17'6"
Closet
Bath
Closet
Bedroom 2 10'8" x 12'0"
Master Bedroom Suite 10'8" x 24'0"
Deck (Optional)
Sunroom (Optional)

Second Floor

Loft 21'4" x 12'0"
Down

Note: Very low headroom

GAMBREL SERIES

SALEM

- 3 Bedrooms
- 2 Bathrooms
- Master bedroom on second floor
- 1762 sq. ft.
- First Floor: 1013 sq. ft.
- Other Floor: 749 sq. ft.
- Overall Size: 27' x 38'

Photos of Gambrels are shown on page 108.

Shading indicates skylight or sunroom.

CONCORD

- 3 Bedrooms
- 2 Bathrooms
- Master bedroom on second floor
- 1291 sq. ft.
- First Floor: 667 sq. ft.
- Other Floor: 624 sq. ft.
- Overall Size: 21' x 31'

LEXINGTON

- 3 Bedrooms
- 1 1/2 Bathrooms
- Master bedroom on second floor
- 1614 sq. ft.
- First Floor: 827 sq. ft.
- Other Floor: 787 sq. ft.
- Overall Size: 27' x 31'

Salem — First Floor: Bedroom 2 10'8" x 11'9"; Closet; Closet; Bedroom 3 10'8" x 11'9"; Entry; Bath; D; W; Pantry; Lin; Closet; Kitchen 10'8" x 9'3"; Up; Dining 10'8" x 9'10"; Living 16'0" x 16'11"; Deck (Optional)

Salem — Second Floor: Master Bedroom Suite 26'8" x 18'0"; Closet; Bath; Down; Loft Area; Cathedral Ceiling; Balcony

Concord — First Floor: Bath; W/D; Lin; Closet; Bedroom 3 10'8" x 14'9"; Kitchen 10'8" x 8'6"; Closet; Up; Dining 10'8" x 8'0"; Living 10'8" x 13'3"; Deck (Optional)

Lexington — First Floor: D; W; Lav; Kitchen 10'8" x 11'0"; Family 16'0" x 13'4"; Closet; Up; Dining 10'8" x 14'1"; Living 16'0" x 14'1"; Deck (Optional)

Lexington — Second Floor: Bedroom 2 13'4" x 11'3"; Bedroom 3 13'4" x 11'3"; Closet; Closet; Bath; Down; Closet; Lin; Closet; Master Bedroom Suite 21'4" x 14'1"; Balcony

Concord — Second Floor: Bedroom 2 21'4" x 14'9"; Lin; Closet; Bath; Down; Lin; Closet; Master Bedroom Suite 21'4" x 10'6"; Balcony

POLE SERIES

QUADRANT

- 3 Bedrooms
- 2 Bathrooms
- 1376 sq. ft.
- Overall Size: 37' x 37'

Photos of Poles are
shown on page 109.

QUADRA

- 3+ Bedrooms
- 2 Bathrooms
- 2304 sq. ft.
- Overall Size: 48' x 48'

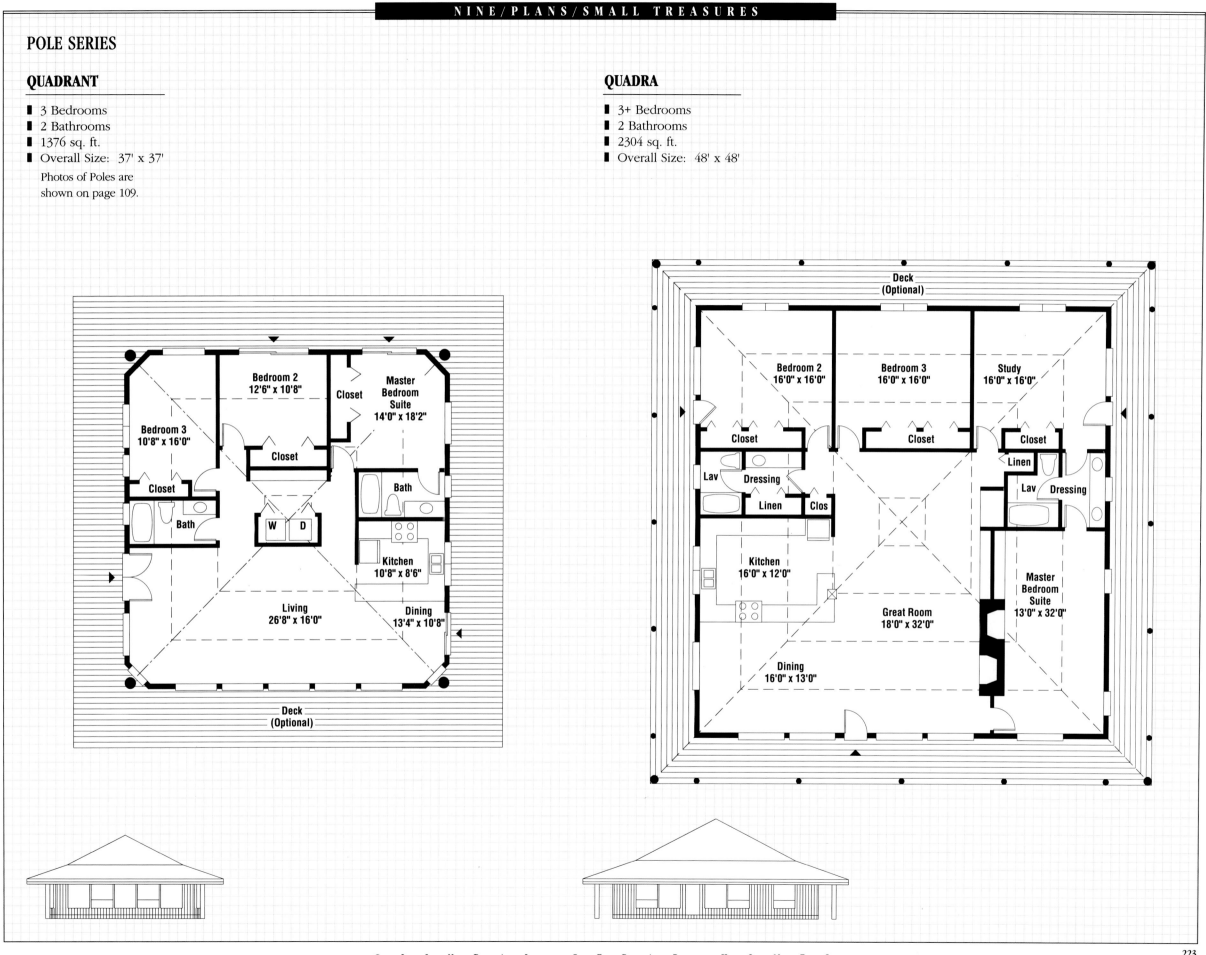

ONE-OF-A-KINDS

	Page Number	First Floor Sq. Ft.	Other Floor Sq. Ft.	Total Sq. Ft.	Number of Floors	Overall Size	Number of Bedrooms	Number of Baths	Master Bedroom on Floor
Romancing the Ranch	P18	3192	982	4174	2	111x60	3	3 & 2	1
Design for Entertaining	P20	2443	735	3178	2	74x51	3+	3	2
Paradise Found	P22	2010	821	2831	2	54x58	3	2	2
New England Light	P24	1506	852	2358	2	53x43	3+	2 1/2	2
Desert Oasis	P25	1843	1236	3079	2	50x48	3+	2 1/2	2
A Stone Cabin Grows Up	P26	3353	1387	4740	2	79x66	4	2 1/2	2
A Modern Log Home	P28	2463		2463	1	98x50	2	2	1
Church	P30	3111	228	3339	2	98x53			
One-of-Kind #9	P115	2155	1189	3344	2	59x56	4	3	2

DESIGNER SERIES

	Page Number	First Floor Sq. Ft.	Other Floor Sq. Ft.	Total Sq. Ft.	Number of Floors	Overall Size	Number of Bedrooms	Number of Baths	Master Bedroom on Floor
Sundowner	P153	1425	591	2016	2	51x33	3	2	2
Sunburst	P154	1902	623	2525	2	60x37	3	2	2
Sunglow	P155	1941	782	2723	2	59x41	3	2 & 2	1

CITY SERIES

	Page Number	First Floor Sq. Ft.	Other Floor Sq. Ft.	Total Sq. Ft.	Number of Floors	Overall Size	Number of Bedrooms	Number of Baths	Master Bedroom on Floor
Boston	P156	910	714	1624	2	35x31	3	2 1/2	2
Toronto	P157	1330	1054	2384	2	43x37	4	3	2
Philadelphia	P158	1439	982	2421	2	43x38	4	3	2

EXECUTIVE SERIES

	Page Number	First Floor Sq. Ft.	Other Floor Sq. Ft.	Total Sq. Ft.	Number of Floors	Overall Size	Number of Bedrooms	Number of Baths	Master Bedroom on Floor
Belair	P159	1220	540	1760	2	47x32	2	2	2
Georgetown	P160	1627	596	2223	2	64x32	3	2 1/2	1
Scarsdale	P161	2466	928	3394	2	68x43	4	3 1/2	1

CONTINENTAL SERIES

	Page Number	First Floor Sq. Ft.	Other Floor Sq. Ft.	Total Sq. Ft.	Number of Floors	Overall Size	Number of Bedrooms	Number of Baths	Master Bedroom on Floor
Tuscany	P162	1035	580	1615	2	37x31	3	2	2
Bavaria	P163	1399	767	2166	2	43x44	2+	2 3/4	2
Burgundy	P164	1450	823	2273	2	43x40	4	3	2

INTERNATIONAL SERIES

	Page Number	First Floor Sq. Ft.	Other Floor Sq. Ft.	Total Sq. Ft.	Number of Floors	Overall Size	Number of Bedrooms	Number of Baths	Master Bedroom on Floor
Madrid	P165	1467		1467	1	43x49	3	2	1
Tokyo	P166	2627		2627	1	66x56	3	2	1

SIGNATURE SERIES

	Page Number	First Floor Sq. Ft.	Other Floor Sq. Ft.	Total Sq. Ft.	Number of Floors	Overall Size	Number of Bedrooms	Number of Baths	Master Bedroom on Floor
Landmark	P167	1216	660	1876	2	43x32	3	2	2
Hallmark	P168	1513	812	2325	2	50x35	3	2 1/2	1

EMBASSY SERIES

	Page Number	First Floor Sq. Ft.	Other Floor Sq. Ft.	Total Sq. Ft.	Number of Floors	Overall Size	Number of Bedrooms	Number of Baths	Master Bedroom on Floor
Consul	P169	1613	1341	2954	2	41x44	3	2 1/2	2
Ambassador	P169	2154	1341	3495	2	56x44	4	3 1/2	1
Diplomat	P170	3148	1272	4420	2	72x51	4	3 1/2	1

TOWN & COUNTRY SERIES

	Page Number	First Floor Sq. Ft.	Other Floor Sq. Ft.	Total Sq. Ft.	Number of Floors	Overall Size	Number of Bedrooms	Number of Baths	Master Bedroom on Floor
Fairway	P171	1461	1365	2826	2	50x32	4+	3	2
Fairlane	P172	1796	1592	3388	2	51x37	4+	3	2

CENTENNIAL SERIES

	Page Number	First Floor Sq. Ft.	Other Floor Sq. Ft.	Total Sq. Ft.	Number of Floors	Overall Size	Number of Bedrooms	Number of Baths	Master Bedroom on Floor
Centurion	P173	1120	1052	2172	2	49x27	2	2 1/2	2
Century	P174	1775	1652	3427	2	61x37	3	2 1/2	2

MASTERPIECE SERIES

	Page Number	First Floor Sq. Ft.	Other Floor Sq. Ft.	Total Sq. Ft.	Number of Floors	Overall Size	Number of Bedrooms	Number of Baths	Master Bedroom on Floor
Rembrandt	P175	1631	476	2107	2	58x42	3	2 1/2	1
Michelangelo	P176	2148	872	3020	2	69x46	4	2 1/2	1
Mozart	P177	2182	1521	3703	2	73x49	4	2 1/2	2

PROW STAR SERIES

	Page Number	First Floor Sq. Ft.	Other Floor Sq. Ft.	Total Sq. Ft.	Number of Floors	Overall Size	Number of Bedrooms	Number of Baths	Master Bedroom on Floor
Venice Vista	P148	1453	381	1834	2	60x36	3	2	2
Bay Vista	P178	1190	197	1387	2	54x27	1	2	1
Nova Vista	P179	1254	493	1747	2	41x41	3	2	2
Buena Vista	P180	1322	437	1759	2	50x35	3	2	2
Cascade Vista	P181	1549	268	1817	2	59x32	2	2	1
Sea Vista	P182	1453	381	1834	2	60x36	3	2	2
Omni Vista	P183	1672	349	2021	2	59x36	3	3	2
Lake Vista	P184	1425	551	1976	2	47x37	2	2	2

PRESIDENTIAL PROW STAR SERIES

	Page Number	First Floor Sq. Ft.	Other Floor Sq. Ft.	Total Sq. Ft.	Number of Floors	Overall Size	Number of Bedrooms	Number of Baths	Master Bedroom on Floor
Lincoln	P185	1279	768	2047	2	46x38	3	2 1/2	2
Jackson	P186	1470	1032	2502	2	49x38	4	2 1/2	2

CHALET STAR SERIES

	Page Number	First Floor Sq. Ft.	Other Floor Sq. Ft.	Total Sq. Ft.	Number of Floors	Overall Size	Number of Bedrooms	Number of Baths	Master Bedroom on Floor
Gemini	P187	1162	304	1466	2	45x31	3	2	2
Capricorn	P188	1104	379	1483	2	40x31	2	2	2
Aquarius	P189	1287	383	1670	2	51x30	3	2	2
Virgo	P190	1359	317	1676	2	48x32	2	1 3/4	1
Libra	P191	1507	334	1841	2	59x34	3	2	1
Zodiac	P192	1303	952	2255	2	48x36	3	2 1/2	2

CONTEMPO PROW STAR SERIES

	Page Number	First Floor Sq. Ft.	Other Floor Sq. Ft.	Total Sq. Ft.	Number of Floors	Overall Size	Number of Bedrooms	Number of Baths	Master Bedroom on Floor
Mercury	P193	1515		1515	1	54x35	3	2	1
Jupiter	P194	1573		1573	1	62x40	2	2	1
Venus	P194	1593		1593	1	57x33	3	2	1
Galaxy	P195	1719	1755	3474	2	74x41	3	2 1/2	1

	Page Number	First Floor Sq. Ft.	Other Floor Sq. Ft.	Total Sq. Ft.	Number of Floors	Overall Size	Number of Bedrooms	Number of Baths	Master Bedroom on Floor
TRADEWINDS SERIES									
Windjammer	P196	1676	366	2042	2	58x50	2	2	1
Windsong	P197	1928	355	2283	2	64x54	3	2	1
HERITAGE SERIES									
Hancock	P198	981	596	1577	2	37x30	3	2	2
Yorktown	P199	1333	815	2148	2	47x34	3+	2 3/4	2
COLONIAL SERIES									
Mayflower	P200	1363	512	1875	2	48x35	3	2	2
Pilgrim	P201	1330	983	2313	2	37x48	3	2 1/2	1
LIBERTY SERIES									
Independence	P202	938	670	1608	2	42x23	3	2	1
Chesapeake	P203	1128	947	2075	2	45x27	3	3	2
FARMHOUSE SERIES									
Pioneer	P204	1067	793	1860	2	40x27	3	2 1/2	1
Homestead	P205	976	743	1719	2	39x30	3	2	1
Harvester	P206	1344	1024	2368	2	42x32	4	2 1/2	1
NEW ENGLAND SERIES									
Patriot	P207	1075	954	2029	2	40x29	3	2	1
Whaler	P208	1152	1141	2293	2	37x32	4	2 1/2	2
Constitution	P209	1551	1503	3054	2	44x39	3+	3	2
PROW SERIES									
Venice	P210	688	328	1016	2	21x34	2	2	2
Viking	P210	1090	786	1876	2	29x42	3	2	2
Vista	P211	790	456	1246	2	21x39	2	1	1
Volare	P211	1019	605	1624	2	27x40	3	2	2
Victory	P212	1461	839	2300	2	32x48	3	2 1/2	1
Vegas	P212	507	259	766	2	21x26	1	1	1
CHALET SERIES									
Matterhorn	P213	608	565	1173	2	21x29	2	1 1/2	2
Grenoble	P213	806	381	1187	2	28x30	2	1	1
Stowe	P214	832	445	1277	2	25x36	2	2	2
Tahoe	P214	1039	561	1600	2	27x40	3	2	2

	Page Number	First Floor Sq. Ft.	Other Floor Sq. Ft.	Total Sq. Ft.	Number of Floors	Overall Size	Number of Bedrooms	Number of Baths	Master Bedroom on Floor
SUMMIT SERIES									
Sierra	P215	913	520	1433	2	27x37	2	1	1
Shasta	P215	1109	379	1488	2	32x37	2	2	2
Teton	P216	1077	399	1476	2	32x36	3	2	2
Olympic	P216	1346	341	1687	2	43x35	3	2	2
CONTEMPO PROW SERIES									
Constellation	P217	1068	258	1326	2	37x32	3	2	2
Neptune	P217	1517	370	1887	2	43x39	3	2	2
Corona	P218	1173		1173	1	32x39	2	1	1
Polaris	P218	1485	347	1832	2	43x39	2+	3	1
VIEW SERIES									
Capri	P219	720		720	1	27x27	2	1	1
Daytona	P219	907		907	1	27x34	2	1	1
Malibu	P219	1016		1016	1	45x27	3	2	1
Riviera	P220	1108		1108	1	29x41	3	2	1
Waikiki	P220	1155	574	1729	2	32x37	2	1 1/2	1
PANORAMA SERIES									
Acapulco	P221	952	232	1184	2	37x26	4	2	2
Monaco	P221	1337	218	1555	2	43x37	3	1 3/4	1
GAMBREL SERIES									
Concord	P222	667	624	1291	2	21x31	3	2	2
Lexington	P222	827	787	1614	2	27x31	3	1 1/2	2
Salem	P222	1013	749	1762	2	27x38	3	2	2
POLE SERIES									
Quadrant	P223	1376		1376	1	37x37	3	2	1
Quadra	P223	2304		2304	1	48x48	3+	2	1
SOLAR SERIES									
Solar #1	P145	1593	1465	3058	2	66x32	4	2 1/2	1

CHAPTER TEN

HOW TO GET YOUR DREAM HOME

With Lindal, your dream is closer than you might think. All it takes to get started is a visit to your Lindal dealer. And chances are that's not far from where you live right now. Because our homes are sold through the largest and most experienced network of  *local dealers in North America. So come on in with your questions, your plans —your dreams. You'll find your Lindal dealer is ready and willing to help you create the most exciting custom cedar home in the world. Your own.*

HERE'S WHAT YOUR DEALER CAN DO FOR YOU

A lot of planning and coordination go into any custom home. So it's reassuring to know that, from your wish list to working blueprints, your local Lindal dealer can provide the service and design expertise you need to transform your dream into reality. That includes making the most of your building site. Giving you the design consultation you need. Analyzing the cost and feasibility of your plan – and helping you get the most home for your money.

Because they live and work in the areas they serve, Lindal dealers can also assist in locating builder contractors and subs, help secure financing, insurance and construction permits. They're well-versed in local zoning and building considerations. They'll make sure you are, too.

Your dealer places your Lindal order and oversees its progress all the way through delivery and inventory. But dealer support doesn't stop there. Consider your dealer a lasting resource for everything from warranty information and maintenance tips to the full range of Lindal products that can add to and enhance your home in the future. We're proud to say that many lasting friendships have evolved out of the experiences our dealers and their homebuyers share.

Some home designs will simply work better than others for you, and your Lindal dealer can tell you why. So take advantage of this opportunity to have your plans analyzed for site suitability, functionality, aesthetics and cost. Your dealer can point out any problem areas and provide hardworking design solutions.

Lindal's local independent dealers are all over the map - 350 home dealers and 125 sunroom dealers around the world. Most are located throughout the United States and Canada, with overseas dealers in Japan, Korea, New Zealand, Australia, the South Seas, Spain, Great Britain, West Germany, and the Carribean.

We've been in the business long enough to recognize the relationship between outstanding local dealers and satisfied customers. So Lindal University gives our dealers the in-depth training and ongoing support they need to be their best. And our customers can count on the same high standards of personal service whether they're in New York or New Zealand – because every Lindal dealer in the world honors a Code of Ethics written by our dealers themselves.

Most local Lindal dealers have a display home for you to explore. So come on in. Roam around inside. Get close to the craftsmanship. And take your time. Sit down, make yourself at home, and imagine what it will be like to have a Lindal home of your own. We encourage you to take advantage of the opportunity to experience Lindal design firsthand and inspect the materials and construction that make us the world's leader in custom cedar homes.

Don't hesitate to ask your Lindal dealer any questions you have about Lindal homes – at any time. In fact, it's a good idea to keep a running list of questions along with your wish list and any drawings you have. Jot them down as they come up so you'll remember them next time you talk with your dealer. If it's important to you, it's important to us

Dear Dave:

Now that our house is finished, I thought this would be a good time to express our thanks. We felt the scale model you constructed helped solve the integration of the roof lines as well as visualize the finished home. Lindal's bundled package system made it easier for me to act as my own general contractor because all components were on site and were engineered for this unit (not to mention the time saved in procurement).

Cathy and I feel we got excellent value for the money spent -- more importantly, we got exactly what we wanted! Our house is now a beautiful home, in which we can be justifiably proud. Being able to physically participate in its construction was very important to us, not only because it saved us money, but because our home is a statement about ourselves.

Though we fully expect this house to far outlast us, we would, without a moment's hesitation, contact you to build another if the need arose.

Sincerely,

Karl & Cathy Rousett

Karl & Cathy Rousett
Eagle Creek, Oregon

If you've enjoyed sketching your home on the grid paper in this book and want to continue the design process yourself, your local dealer will be happy to give you a Lindal Home Planning Kit -along with helpful tips on making the most of it. Inside you'll find a home planning guide and checklist that walk you through important aspects of planning, furniture cutouts you can trace onto your plan, and a cedar pencil. Most important, this kit contains lots of 1/4" grid paper and a design template just like the one used by our designers at Lindal headquarters. These tools let you take your plan up to 1/4" scale, the actual scale used in construction plans. And your kit is free when you visit your local Lindal dealer.

THERE'S A LOT TO LOVE ABOUT YOUR LINDAL

*T*ime and again, Lindal homeowners tell us they chose Lindal for our clear understanding of their wants and needs – and for our ability to meet them. So what is it that sets us, and our homes, apart? Here's why Lindal homeowners say they chose Lindal – and why, years later, they are happy they did:

- The style and appearance of Lindal home design.
- Quality that lasts - inside and out.
- The service and support of local Lindal dealers, from start to finish.
- The benefits of our experience – over nearly five decades and tens of thousands of homes.
- Our size and financial strength, which allow us to pass the savings of volume buying on to you, and to operate our own sawmill for guaranteed quality control.
- Cost-effective worldwide delivery, resulting in more home for your money.
- Our guarantee – the best in the industry.

Ultimately, what makes all of the above possible is our pride in your success. Although we're the world's largest manufacturer of cedar homes and a publicly-owned company, Lindal always has been – and always will be – driven by a family spirit that takes pride in the knowledge that pursuing our dream is helping you to achieve yours.

What Will Your Lindal Cost?

Because each and every Lindal home is unique, your costs will depend on many of the design decisions and specification choices you make. This flexibility is one of the beauties of building a Lindal. But it means that you need to ask yourself some questions before your dealer can answer the one about cost. For example, how many bedrooms and baths do you want? Do you envision a one- or two-story home? That's important, since two-story homes tend to cost less than one-story homes with the same square footage.

Once you have a sense of overall design, how about Lindal specifications? Which style do you prefer: a Lindal, Justus, Clapboard or Round Log home? Which floorplan, if any, do you want to work from? Your choice of optional materials can affect cost significantly: asphalt roof shingles cost much less than cedar shakes, and drywall is less expensive than wood paneling. Keep in mind that site and local labor costs have a significant impact on construction costs. Naturally, what you put inside your home – from appliances to floor coverings – will influence the total price, too.

In getting a handle on cost, your Lindal dealer can help you to ask the right questions and plan the most home for your money. And once you've made your personal choices about design and specifications, your local dealer can tell you exactly what it costs to move up to your Lindal.

A word of advice as you proceed with planning and pricing: Don't scrimp on quality in the primary materials in your home. Much of your total building budget remains constant: your site, foundation, interior finishing and landscaping. High-quality materials are well worth the relatively small difference they make in the overall cost of your project, both in the pleasure they'll give you and in the value they add to your home.

As a publicly-held company and the world's largest manufacturer of custom cedar homes, we buy our materials in high volumes at low prices, and we pass the savings on to our homebuyers. Our size also allows us to own and operate our own sawmill, which assures a level of quality control that only Lindal can guarantee from the forest to your building site. Finally, our size gives us the financial strength and stability to continually research and develop new products that benefit you and your Lindal home.

Lindal's location in cedar country, where the world's cedar prices are set, allows us to ship virtually anywhere at prices competitive with less substantial, locally-built houses. Our kiln-dried, precision-cut cedar homes mean you don't pay for the expense of shipping oversized, moisture-heavy timber to a local builder. The result? More home for your money.

When your Lindal home is delivered, you'll find it contains only the highest quality building materials custom-engineered and milled to your plan's most exacting specifications. But something else goes into your delivery – our decades of experience and innovation in making the entire process clear, straightforward and streamlined.

Critical building components are part-numbered for easy reference to your blueprints and materials list. And the floor system, wall planks and roof boards are tongued and grooved for easy assembly.

Some people assume that custom homes and manufactured homes are worlds apart. And, in truth, they can be. But not at Lindal. We're known for our unique ability to give our customers the best of both worlds: the highly individual character of a custom home, along with the personal control you maintain when you work from a rich resource base of successful designs.

How many ways can you improve on a dream? We're sure we haven't found all of them yet, but our line of home products keeps growing in our dedication to make your Lindal a growing pleasure to live in. Just a few of the options you can add any time: Lindal Cedar SunRooms, decks, garages, hardwood flooring, our wide world of windows and a grand selection of glass doors. Whether you're adding a second deck or installing a skylight, you'll find Lindal home products are built to match the quality materials and design of your Lindal custom cedar home.

You receive seven comprehensive sets of working blueprints and specifications that include your home's elevations, floor plans, foundation, roof and standard or custom details. Your final Lindal building materials list is computerized and cross-referenced to your final blueprints.

1-800-426-0536

For the name of your local dealer, just call this toll-free 800 number. After you become a Lindal homebuyer, we'll immediately send you – along with your warranty claim number and Homeowner's Kit – a toll-free number reserved exclusively for homeowners. Answering this toll-free number are our professionally trained and caring Customer Services people, who will be happy to answer your questions and handle any warranty claims.

FROM OUR FAMILY TO YOURS

*I*t's good to know there are still a few things in life you can count on. A Lindal home is one of them. In a world where quality has become an everyday promise but a rare practice, people appreciate that it still has deep meaning in a Lindal home. In fact, we're confident that Lindal quality is the best money can buy. And we're proud to share our confidence with you. So not only do you have our word, you have a 10-year guarantee —from our family to yours. And getting started is an easy and exciting one-step process: visit your local Lindal dealer.